Chapter One

General Introduction and Literature Survey

I0476175

1.1. Beta-2-microglobulin

Beta-2-microglobulin (β_2M) is a low molecular weight protein (11.8KD) which was first isolated from the urine of human patients with Wilson's disease, chronic cadmium poisoning, and chronic kidney tubular dysfunction by Beggared and Bearn in 1968[1] .It has since been found in a variety of physiological fluids as well as on the plasma membranes of all nucleated cells as the light chain of the MHC class I antigens of man (HLA) and other vertebrates[2,3].

A non-covalent linkage with the HLA-A, B and C antigens has been demonstrated[4], and β_2M has been shown to be involved in the maintaining the three-dimensional configuration of these antigens[5]. Besides this close association with HLA class I molecules, β_2M has a substantial amino acid sequence homology with immunoglobulin-G constant domain regions[6].

β_2M in serum is derived from HLA class I turnover and shows little day-to-day variation in its concentration in healthy individuals. Serum β_2M values tend to increase with age owing to gradual reduction of the efficiency of its clearance from the blood by glomerular filtration[7].

1

1.1.1. Sources and Distribution of β_2M

β_2M is eliminated from the blood in the kidneys apparently mainly by glomerular filtration followed by tubular resorption and catabolism [8]. Normal humans excrete only about 100mg of β_2M per 24-hour urine volume [9], whereas patients with tubular resorption defects can excrete at least 1000 times as much[9] .Therefore, urine from those patients has been the best source for the isolation of this protein. Materials for initial work on β_2M were obtained from patients with Wilson's disease [1]. But patients with renal failure are another suitable and often more easily available group. They have highly increased levels of B_2M in the blood [9]. And it has become apparent that they also have a highly urinary excretion of the protein [10].

β_2M is normally present in low concentrations in serum [11], saliva [12] and cerebrospinal fluid [13], and in relatively higher concentrations in colostrum [12].

The realization of the close resemblance between β_2M and immunoglobulin domains led to the demonstration that β_2M is present on the plasma membranes of lymphocytes [14]. Bernier and Fanger [15] have reported that short-term lymphocyte cultures from normal human donors released β_2M into medium and that both B and T lymphocytes carry the protein on their membranes. Subsequent works showed that β_2M is also present on the plasma membranes of several types of cells [16]. So far, only two cell types that lack β_2M have been found; namely erythrocytes [17] and the Daudi lymphoma line [18] and accordingly, β2M is synthesized by and located on all normal nucleated cells.

2

β_2M can also be found on sperm [19]. In ontogenetic studies it has been shown that fetal liver, kidney, thymus and testis are also capable of active de novo synthesis of β_2M. On the cell surface, β_2M can be found in free form [20], in association with MHC antigens [21], non H-2 antigens [22] and possibly tumor antigens [23].

1.1.2. Genetics of β_2M

The most striking features of β_2M are its similarity to the homology regions of immunoglobulins and its production by and appearance on nearly every cell type. The internal homologies within the immunoglobulins have suggested that these molecules evolved by duplication of a precursor gene sufficient in size to specify a polypeptide of about 100 residues [24].

β_2M resembles these homology regions to almost the same extent as they resemble each other, suggesting that the gene specifying β_2M evolved from this same gene [24].

Proof that β_2M is an integral chain of human leukocyte antigens (HLA-A, B and C) [25] has provided significant support for the hypothesis that β_2M may be derived from a gene that gave rise to HLA molecules which in turn were precursors of the immunoglobulins [26].

Cell-hybridization experiments [27] indicate, however, that the gene specifying β_2M is not on the same chromosome as the gene for the HLA molecules. It has been reported [27,28] that the gene coding for β_2M is carried on chromosome 15, not on the major histocompatibility complex (MHC) region on chromosome 6 along with HLA.

Gussow D. et al [29] have reported the nucleotide sequence of the human $\beta_2 M$ gene and of nearly full length cDNA clone, the transcriptional start site of the human $\beta_2 M$ gene has also been determined [29].

1.1.3.Structure of $\beta_2 M$

Primary Structure

Unlike many other cell surface proteins, $\beta_2 M$ contains no carbohydrates in its structure, so the 100 amino acid residues account for the entire molecule, which has a molecular weight of 11.8 KD [30]. The amino acid composition of the human $\beta_2 M$ is shown in table (1-1).

Table (1.1). Amino acid composition of the human $\beta_2 M$ [30].

Amino acid	Number	Amino acid	Number
Aspartic acid	8	Glutamine	3
Threonine	5	Methionine	1
Serine	10	Isoleucine	5
Glutamic acid	8	Leucine	7
Proline	5	Tyrosine	6
Glycine	3	Phenylalanine	5
Alanine	2	Lysine	8
Cysteine	2	Histidine	4
Valine	7	Arginine	5
Aspargine	4	Tryptophan	2

The complete amino acid sequence of β_2M has been determined [31] (figure 1-1). It has been shown that the two cystienyl residues form an intrachain disulfide bond. The single methionyl residue is located at the carboxyl terminus of the chain. There is a cluster of aromatic residues in the region of residues 60-71, where the only charged residue is the glutamyl residue at position 70.

The amino terminal portion of the chain (residues 1-13) is basic, and there is a predominance of charged residues in the region of residues 30-53. The acidic residues at positions 47and 50 and 75,77and 78 apparently reduce the ability of trypsin to cleave the lysyl residues at positions 48 and 76. In contrast, the Tyr-Ser bond at positions 10-11 is unusually susceptible to cleavage by chymotrypsin [32].

Although the sequence studies were carried out on pooled samples of human β_2M obtained from a number of unrelated patients, no evidence has been reported neither for any variations in the amino acid residues seen at each position, nor for micro heterogeneity in the molecule. While the presence of genetic variability in the structural gene for this protein cannot be excluded on the basis of these results, it is unlikely that there is a high degree of variation from individual to individual [32].

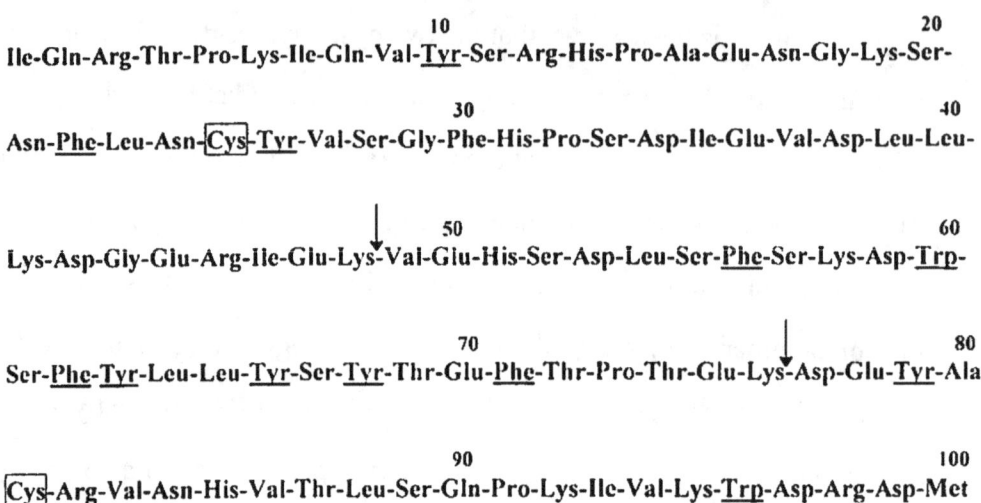

Ile-Gln-Arg-Thr-Pro-Lys-Ile-Gln-Val-<u>Tyr</u>-Ser-Arg-His-Pro-Ala-Glu-Asn-Gly-Lys-Ser-

Asn-<u>Phe</u>-Leu-Asn-Cys-<u>Tyr</u>-Val-Ser-Gly-Phe-His-Pro-Ser-Asp-Ile-Glu-Val-Asp-Leu-Leu-

Lys-Asp-Gly-Glu-Arg-Ile-Glu-Lys-Val-Glu-His-Ser-Asp-Leu-Ser-<u>Phe</u>-Ser-Lys-Asp-<u>Trp</u>-

Ser-<u>Phe</u>-<u>Tyr</u>-Leu-Leu-<u>Tyr</u>-Ser-<u>Tyr</u>-Thr-Glu-<u>Phe</u>-Thr-Pro-Thr-Glu-Lys-Asp-Glu-<u>Tyr</u>-Ala-

Cys-Arg-Val-Asn-His-Val-Thr-Leu-Ser-Gln-Pro-Lys-Ile-Val-Lys-<u>Trp</u>-Asp-Arg-Asp-Met

Figure (1.1). Amino acid sequence of the human β₂M. Half-cystinyl residues are enclosed in boxes and residues with aromatic side chains are underlined. The arrows denote bonds that are resistant to cleavage by trypsin [32].

Secondary and Tertiary Structure

The three-dimensional structure of β_2M is strikingly similar to the structures of the known immunoglobulin constant domains. Employing X-ray crystallographic studies, it has been shown that the molecule is approximately $45 \times 25 \times 20$ A° in size [33].

Almost half of the amino acid residues are involved in a portion of the polypeptide chain which is folded into a typical "β-barrel" configuration dominated by two large anti-parallel β-pleated sheets, one of four strands and the other of three. The four-stranded sheet contains residues 6-11, 28-21, 61-69,and 56-49; the three-stranded sheet contains residues 90-94, 82-77, and 36-40. The single intrachain disulfide bond connects the two β-structures forming a loop of 57 residues in the polypeptide chain [33]. No significant amount of α–helical structure has been reported [34].

It has also been shown that most amino acid residues that are the same in the sequences of β_2M and immunoglobulin constant domains have their side chins in the interior of the molecule between the β sheets [33].

Hydrodynamic studies have previously shown β_2M to be nearly spherical and highly compact. Consistent with this view of the molecule is the fact that under native conditions it is resistant to proteolysis [35]. In crystals studied, β_2M is clearly monomeric and apparently does not form dimers in vivo but associates with the heavy chains of human leukocyte antigens [33,36].

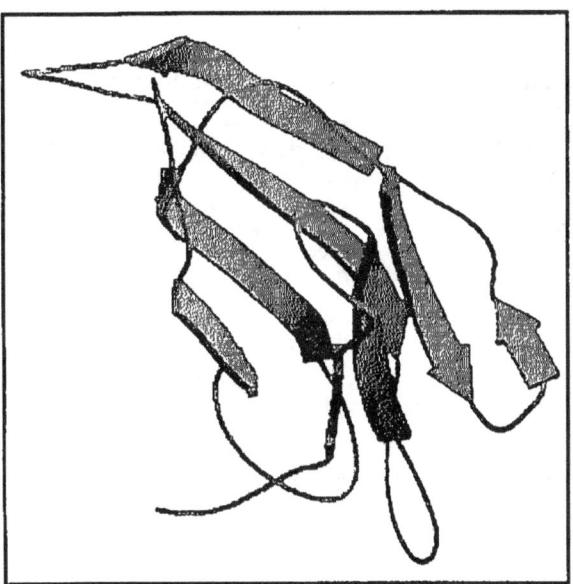

Figure (1.2). The three-dimensional structure of β_2M. The two anti-parallel β-pleated sheets are shown as large arrows [37].

Association with HLA (Quaternary structure)

Human leukocyte antigens (HLA) are a set of highly polymorphic glycoproteins expressed at the surface of all nucleated cells [38]. As the light chain of HLA molecule, $\beta_2 M$ has been shown to be non-covalently attached to the heavy chain which consists of three extra cellular globular domains designated α_1 (N-terminal), α_2 and α_3, transmembrane region and a cytoplasmic tail [39]. The interaction of $\beta_2 M$ with the heavy chain has long been thought to be mediated mainly by the third (α_3) domain [40].

Rein R.S. et al.[41] have reported that also the first (α_1) and the second (α_2) domains of the heavy chain make an essential contribution to subunit interaction.

On the basis of the crystal structures of human leukocyte antigens, Tysoe-Calnon V.A. etal [42] have shown four distinct regions form the contact points between the heavy chain and $\beta_2 M$, one on each of the α_1, and α_2 domain and two on the α_3 domain.

Evidence from solution studies has suggested that the four-stranded β-sheet of $\beta_2 M$ may be involved in the association with HLA heavy chain domains [43]. A monoclonal antibody is known to bind to Arg-45 of human $\beta_2 M$ both when the molecule is free and when it is complexed with HLA class I heavy chain [44]. This binding would leave the four-stranded β-sheet free to interact with the heavy chain domains [44].

It has also been observed that the tyrosine residues 10, 26 and 63 of human $\beta_2 M$ are iodinated in the free molecule but not when it is attached to HLA heavy chain [45]. In the three-dimensional structure, these protected side chains are exposed on the surface of the four-stranded beta structure where they would be in non-covalent contact with the HLA class I molecules heavy chain domains [33,46].

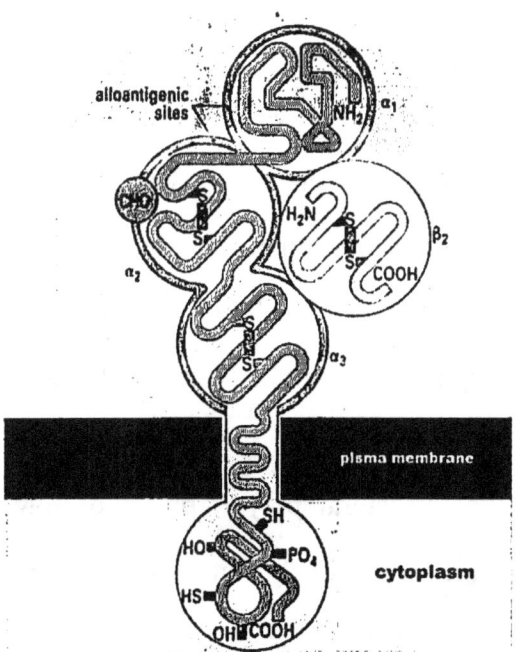

Figure (1.3). Association of β_2M with HLA class I molecule. The globular domains (α_1 , α_2 and α_3) are shown in deep grey. The α_3 domain is closely associated with the non-MHC encoded peptide β_2M, which is shown in light grey [47].

1.1.4.Function of β_2M

In many ways, β_2M appears to be a molecule of the immune system. Whether B_2M has a primary immunological role is at present uncertain, but interest about its function has been stimulated by the knowledge that HLA system is involed in the regulation of many immure responses including those concerned with the recognition of self and non-self (antigen presenting to cytotoxicT-cells)[48], and non-immunological functions associated with hormone-growth factor receptor interactions [49] and cell proliferation [50].

It has been shown that the presence of β_2M is necessary for HLA heavy chain to be post translationally processed and properly expressed on the cell surface [51,52], and that viral antigen that block the association between the two chains can prevent the expression of HLA molecules on the cell surface [53,54]. Also, mice lacking a functional β_2M gene do not express class I molecules on the cell surface and have defective cell-mediated cytotoxicity [55].

The non-covalent association of β_2M to the heavy chain is essential for the structural stability and optimal function of HLA class I molecules [56]. The beta sheet regions that dominate the structure of β_2M are believed to contribute in providing a supporting floor of β-strands for the antigene binding cleft on HLA class I molecule [57].

The lack of structural polymorphism of β_2M argues against a direct receptor function, it is more likely that β_2M exerts a receptor-coordinating effect on the various receptor systems [58]. β_2M has also been reported to have an important role in T-cell precursor colonization in the thymus [59,60].

Mori M. etal. [61] have demonstrated that β_2M plays an important role in regulating the elimination of tumor cells which occurs as a result of the action of β_2M as an apoptosis-inducing factor.

1.2. β₂M as a Tumor Marker
1.2.1. Tumor Markers

The result of transformation is a malignant cell that in each cycle of cell division generates a new malignant cell. In this process, the malignantly transformed cells acquire some new properties through which they differ from malignant cells of the same origin. The acquired properties can be either the changes in cellular morphology, physiology, or the changes in cell behavior [62].

These subtle differences between normal and malignant cells are therefore being exploited in the detection of malignancy, and the substances that are being determined in this process are termed tumor markers. Different fields in oncology utilize different tumor markers according to their need and their techniques of follow up. Theoretically, the possibilities for the application of tumor markers in oncology are numerous, but the utilization depends upon the sensitivity and specificity of the marker and upon reliability of other methods that are being used for the same purpose [63].

Clinically important utilization of markers includes [63]:

(i) Early detection of the tumor.

(ii) Differentiating benign from malignant conditions.

(iii) Evaluating the extent of the disease.

(iv) Monitoring the response of tumor to therapy.

(v) Detecting the recurrence of the tumor.

Up till now, no antigenic structure is known that would be present only in tumor cells and that means that antibodies against certain tumor markers cross react also with other antigenic structures. Therefore, it has been concluded that no tumor marker and no methods of following up the presence of malignant cells are 100% specific. When assessing these results, it has been found that not only a malignant disease causes elevated levels of tumor markers, but there are also other factors that effect their concentration the most common incompetence's of tumor markers are inadequate specificity for the type of malignancy, production of markers in high concentration in nonmalignant disease, production of markers in different physiological conditions and production in perfectly healthy tissues [64]. Tumor markers can be classified in several ways, according to their chemical structure, their tissues of origin, type of malignancies in which they are elevated, etc. The most common classification tries to combine their biochemical properties, tissue of origin and functionality, according, they are classified as [63]:

    (i)   Oncofetal proteins (*e.g.* CEA).

    (ii) Hormones (*e.g.* hCG).

    (iii) Enzymes (*e.g.* NSE).

    (iv) Tumor-associated antigens (*e.g.* CA125).

    (v)   Miscellaneous markers (*e.g.* TPA).

    (vi) Special serum proteins (*e.g.* β_2M).

1.2.2. Relevance of β_2M to Tumor

There have been many attempts to identify cell surface structures unique to malignant cells. Much evidence has been collected to show that malignant cell transformation involves changes in surface charge density, loss or gain of antigens, and loss of contact inhibition [65]. Consequently, understanding the structure of both normal and abnormal cell membranes became the dominant occupation of researchers during the past decades [65,66,67].

β_2M, being a component of several surface antigens, is considered a potential marker of malignancy [68]. Studies on a variety of cell lines have been shown that there are some tumors, which fail to express β_2M and /or HLA class I molecules at least in vitro. The Daudi cell line synthesizes HLA antigens but these cannot proceed to appear on the cell surface since there is an absence of β_2M [69]. Whereas cell lines from a choriocarcinoma secrete β_2M in the absence of HLA class I molecules so it could be expected that there would be increased levels of circulating free β_2M with such tumors [70]. In melanoma, whose tumor cells often lack HLA molecules; distinct mutations in the β_2M gene have been identified [71]. Expression of β_2M can therefore be used as a marker for the presence of HLA class I molecules [72].

The loss and the aberration of the expression of β_2M and /or HLA in many solid tumors have been extensively studied [73,74]. These abnormalities in the mechanisms of expression have always been regarded as one of the pathways by which tumor cell evades from the destructive immune responses and has created problems in the design of immunotherapy [75,76]. Partial or complete loss of HLA class I molecules has been described in several human tumor types [76,77] and associated with unfavorable factors such as lack of differentiation [78,79], invasiveness [80] and metastatic potential [81].

1.2.3.β_2M in Cancer

The close correlation between serum levels of β_2M and glomerular filtration rate is the basis for the clinical use of β_2M determination [82]. In evaluating a possible role of circulating β_2M as an indicator of malignant disease, critical point is the adequacy of controls. This problem is further emphasized by the age, dependence of β_2M levels reported earlier [82].

Increased serum levels of β_2M have been reported in a number of malignant condition including breast [83], lung [84], prostate [85], ovarian [86], Oral cancer [87], urothelial [88], neuroblastoma [89] and renal cell carcinoma [90].

β_2M is clinically used for lymphoproliferative diseases including leukemia [91], lymphoma [92] and multiple myeloma, where serum β_2M is related to tumor cell load, prognosis and disease activity [93].

The mechanism of increase in β_2M levels in malignancies is not known but various possible hypotheses for the increase have been put forward. The β_2M is a cell membrane constituent, so an accelerated membrane turnover or accelerated cell division could increase the shedding of β_2M [94]. The ability of the carcinoma cells to produce a higher concentration of β_2M than the non-neoplastic cells may be due to either active synthesis or increased cell breakdown or both [95].

β_2M elevations are not diagnostic of cancer as a number of non-malignant conditions also give rise to elevated concentrations [96,97]. Although β_2M levels are elevated in some solid tumors, the marker is not useful in prognosis or in monitoring the disease state in these situations [98].

1.3.Ligand-Receptor Interactions

1.3.1.General View on Protein Interactions

Proteins are responsible for a large variety of tasks in biological systems, such as enzymatic catalysis of chemical reactions in metabolism, defence against infection, regulation and transport .For all these tasks highly specific or, in other cases, more general interactions with other molecules are necessary. Proteins achieve these by different surface motifs and a multitude of structural arrangement in the molecular interface [99].

The elucidation of bimolecular interactions is of steadily increasing importance. Exact knowledge of the principles governing the strengths and formation of molecular interactions is of highest importance for a huge number of applications in widely different areas [99].

The experimental determination, evaluation and /or prediction of protien interactions with other proteins, small molecules, sugars or DNA is a requirement for [99]:

1. The design of new drugs.
2. The design of proteins stable in different environment.
3. Understanding the cross reactions of antibodies used in medical diagnosis or treatment.
4. The prediction of possible side effects of drugs interacting with more than one protein.
5. The improvement of our understanding in disease like Alzheimer's disease.

6. Understanding and regulation of biocatalytic activity.

7. Understanding cell-cell communication and cell differentiation.

1.3.2. Nature of Interaction Forces

Many studies have been made to clarify the forces that determine ligand-receptor interactions and how these forces are used to yield complexes with association constants that vary over many orders of magnitude and that can be controlled by deferent exogenous agents [100].

Structural studies on proteins have showed that a significant number of charged and polar residues, which predominate protein surfaces, are buried upon complex formation. Interactions might be non-polar, hence, to be similar to the interiors of the individual monomer, where hydrophobicity is an average characteristic property on interaction surfaces [101].

The non-covalent forces that hold the tow proteins forming the complex include hydrogen bonds, electrostatic, dispersion and hydrophobic forces. Among these, electrostatic forces are the strongest and the most important factors in the interaction. Although hydrogen bonds are weaker than electrostatic forces, their numbers make them contributors of interest in stabilizing the complex. The weakest contribution to binding strength is provided by Van der Waals forces. Hydrophobic interactions function by the exclusion of polar water molecules to bring hydrophobic molecules together, such interactions also serve to attract polar water molecules to hydrophilic amino acid residues on the interacting protein molecules [102].

1.3.3.Physical Perspective on Ligand - Receptor Binding

Most analysis of radioligand binding experiments are based on a simple model called the law of mass action that is based on these simple ideas [103]:

- All receptors are equally accessible to ligands and binding occurs when ligand and receptor collide due to diffusion when it has the correct orientation and enough energy. The rate of association equals [ligand].[Receptor]. k_{on}, where k_{on} is the association rate constant in units of $M^{-1}min^{-1}$.

- All receptors are either free or bound; the model ignores any states of partial binding. The ligand and receptor remain bound together for a random amount of time influenced by the affinity to each other .The rate of dissociation equals [ligand-Receptor]. k_{off} , where k_{off} is the dissociation rate constant expressed in unites of min^{-1}.

- Neither ligand nor receptor is altered by binding, after dissociation, they are the same as they were before binding.

- Binding is reversible and reaches equilibrium when the rate at which new ligand-receptor complexes are formed equals the rate at which the ligand-receptor complexes dissociate.

At equilibrium:

$$[Ligand].[Receptor]. k_{on} = [Ligand-Receptor] k_{off}$$

$$\frac{[Ligand].[Receptor]}{[Ligand - Receptor]} = \frac{k_{off}}{k_{on}} = K_d$$

The dissociation constant (K_d) expressed in units of molar, is the concentration of ligand that occupies half of the receptors at equilibrium. A small K_d means that the receptor has a high affinity for the ligand, and a large K_d means that the receptor has a low affinity for the ligand. Radioligand binding reaction can also be expressed by the inverse of (K_d), the association constant (K_a), which is useful in determining the strength of the binding [103].

1.3.4. β₂M as a Radioligand

A radioligand is a radio actively labeled molecule that can be bound to a receptor or any protein of interest. Charnet A. etal.[104] have described β_2M as a ligand of α_2-macroglobulin and identified the regions of the sequence of β_2M that have the capacity to bind to α_2M. Also, the radiolabeled β_2M has previously been shown to have the cryophilic property to be bound to cell membrane components (receptors) of a number of species [105].

Working on murine cells, Sege K.etal [106] have demonstrated a cell surface component with affinity for exogenously added ^{125}I-β_2M. They have shown that the "receptor" binds β_2M in a reversible manner and does not seem to express any of the alloantigenic determinants specifying the classical H-2K and D antigens.

Although the term "receptor" is commonly used to express proteins that upon their binding activate a particular biological process, in this study it has been used for brevity and does not imply this meaning.

1.4.Breast Tumors

The breast is a bilateral organ that in the female undergoes dramatic changes in size, shape and function in association with infantile growth, puberty, pregnancy, lactation and post-menopausal regression[107].

The breast is also the source of the most frequently diagnosed malignancy in female population, breast tumors account for up to one third of all new cases of women's cancer[108]. It is a truism to suggest that breast tumor is the result of a subtle imbalance in the complex regulatory cycles to which breast tissue is exposed. Changes in hormonal, nutritional, genetic, physiological and environmental stimuli such as radiation that influence normal growth and function do so by up and down regulating genetic pathways leading to cell proliferation and regression[109].

The development of breast cancer has been postulated to be a multistep process that follows a defined sequence of qualitatively different events, namely progression of ductal hyperplasia and atypical ductal hyperplasia, which represent the initial stages of neoplastic growth, to carcinoma *in situ* , invasive carcinoma and ultimately metastasis , as has been documented for a number of other malignancies[110].

It is not known when in the lifetime of a woman breast tumor initiates.The term tumor is applied indistinctly to either benign or malignant lesions[111] table (1.2) summarizes breast tumor classification. Studies of chemically induced carcinogenesis in an experimental animal model and primary cultures of human breast epithelial cells have shown that the initiation of the neoplastic process is inversely related to the degree of differentiation and *in vivo* cell proliferation of the mammary gland [112]. An important concept that emerges from these studies is that all cancers are related to the terminal duct lobular unit [113].

Breast tumors are age related because multiple defects are involved; the clinical stage and pattern of disease are function of these defects. Moreover, there is a dynamic relationship between tumor and the host, and thus the defects are themselves a constant state of flux [114].

The risk of developing breast cancer increases with age, the disease is uncommon in women under the age of 40 years; only about 0.8% of breast cancer occur in women under age 30 and approximately 6.5% in women between 30 and 40 years old [109].

Table (1.2). Classification of breast tumors

I. Epithelial tumors
A. Benign
1. Intraductal papilloma
2. Adenoma of the nipple
3. Adenoma
a. Tubular
b. Lactating
B. Malignant
1. Non invasive
a. Intraductal carcinoma
b. Lobular carcinoma
2. Invasive
a. Invasive ductal carcinoma
b. Invasive lobular carcinoma
II. Tumors of Connective and epithelial tissues
1. Fibroadenoma
2. Pyllodes tumor
3. Carcinosarcoma
III. Miscellaneous Tumors
1. Soft tissue tumors
2. Skin tumors
3. Tumors of lymph tissues
IV. Mammary Dysplasia
V. Lesion Like Tumor
1. Ductectasia
2. Inflammatory pseudo tumors

1.5. Aim of the work

1. Determination of β_2M concentrations in sera of patients with benign and malignant breast tumors.

2. Measurement of β_2M receptors in tissues of benign and malignant breast tumors.

3. Characterization of β_2M binding to its receptors in benign and malignant breast tumors and studying different factors affecting the binding reaction such as ^{125}I-β_2M concentration, protein concentration, pH, temperature and time of reaction.

4. Determination and study of the kinetic and thermodynamic parameters of ^{125}I-β_2M binding to its receptors in benign and malignant breast tumors.

5. Spectroscopic studies of β_2M and (^{125}I-β_2M/receptor) complex using ultraviolet spectroscopy.

CHAPTER 2

Preliminary Studies on $^{125}I\text{-}\beta_2M$ Binding to its Receptors

Chapter Two

^{125}I-β_2M Binding to its Receptors Preliminary Studies

Introduction

Beta-2-microglobulin is a single-chain aglycosyl protein composed of 100 amino acids. Its molecular mass is 11.8 KD, and it is now known to be the light-chain component of the HLA class I molecules. It is therefore found on all nucleated cells and is present in high concentration on the lymphocyte cell surface[115].

Increased serum levels of β_2M have been reported in malignancies of B-cell origin, such as multiple myeloma[116]. In solid tumors, moderate elevations of serum β_2M have been observed indicating that it may be of use as a tumor marker.

There have been conflicting reports on serum β_2M in breast cancer patients. Klien etal[83] have reported a significant increase in β_2M in breast cancer patients compared with normal controls. However, in a study of 129 breast cancer patients, Adami etal[117] have shown that there were no significant differences in β_2M levels compared with age-matched controls. Hence, this part of the study was undertaken to evaluate further serum levels of β_2M.

In considering the binding studies, which deal with β_2M as a ligand, there has been some confusion concerning the ability of exogenously added β_2M to bind to cell membrane components of human[105]. It appeared of importance to physically describe the binding process. A necessary precursor for this characterization is determination of the most appropriate environment for the process using the radioimmunoassay technique with our modification for tissue as no simple system could meet all requirements for this aim.

Materials and Methods

2.1. Materials
2.1.1. Chemicals

All chemicals and reagents used in this study were of analar grade and were used without further purification. There specifications are summarized in table (2.1)

Table (2.1). Specifications of Chemicals Used.

Chemicals	% Purity	Company
RIA kit for β_2M	-	Immunotech.(Czeh Republic)
BSA	99	Fluka (Switzerland)
$ZnCl_2$,NiCl,$CuCl_2$,$MnCl_2$	99	Fluka (Switzerland)
NaF,NaCl,NaBr,NaI	99	Fluka (Switzerland)
$CuSO_4$, Na,K-tartrate ,EDTA	98	Fluka (Switzerland)
NaOH, HCl, $CuSO_4$	98	BDH (U.K.)
PEG 6000 ,tris(hydroxymethyl)aminomethane,and sucrose	99	BDH (U.K.)
Folin ciocalteau	99	E.Merck AG.

2.1.2. Instruments

All instruments used in this study are listed in table (2.2).

Table (2.2). Instruments and their manufacturer companies.

Instruments	Company
Gamma counter type 1270 rack gamma II	LKB
Double beam spectrophotometer	Shimadzu
pH-Meter	Pye Unicam
Cooling centrifuge (max. 5000 r.p.m)	MSE
Memmert water bath ,memmert incubator	Gallenkamp
SM- Shaker	Barnson

2.1.3. Patients

Seventy-five patients were included in this study. They were divided into three groups, the first one (GI) consisted of post-menopausal women with malignant breast tumor, the second group (GII) consisted of pre-menopausal women with malignant breast tumor and the third group (GIII) consisted of benign breast tumor patients. A forth group of normal women was also included as a control group. The host information for all patients and healthy women are summarized in table (2.3).

Table (2.3). The host information of patients and normal individuals included in this study.

Group	Patients	No.	Age-range	Type of tumor
I	Post-menopausal malignant breast tumor	28	55-70	Infiltrative ductal carcinoma
II	Pre-menopausal malignant breast tumor	22	23-48	Infiltrative ductal carcinoma
III	Benign breast tumor	25	19-50	Fibroadenoma
IV	Healthy individuals	20	23-36	-

All patients were admitted for diagnosis and/or treatment to Medical City Hospital, Nursing Home private Hospital, Al-Yarmuk Teaching Hospital and Al-Jadriyah Private Hospital. All surgical operations were carried out under the supervision of surgeons. To avoid any interfering, patients who suffering from other diseases were excluded.

2.2. Methods
2.2.1. Blood Sampling

Three to five milliliters of blood samples were obtained by vein puncture just before surgery, the whole blood was left for 20 minutes at room temperature, after coagulation the serum was separated by centrifugation at

3000 r.p.m for 10 minutes, then the serum samples were aspirated and kept frozen at -20 °C until assaying.

2.2.2.Tissue Collection

Three samples of breast tumor tissues were included in this study; they were surgically removed from a post-menopausal patient with malignant breast tumor, a pre-menopausal patient with malignant breast tumor, and a patient with benign breast tumor. The obtained tumor tissues were immediately rinsed with ice-cold isotonic saline solution; they were individually collected and kept at -20 °C until homogenization.

2.2.3. Preparation of Tris-Buffer Solution

Tris-buffer solution of total concentration 0.05M and pH 7.2 was prepared by dissolving 0.6075 gm of tris(hydroxymethyl)amino methane,20 ml glycerol, 8.55 gm of sucrose and 0.1816 gm of EDTA in 100ml of deionized water then the pH was adjusted to 7.2 with concentrated hydrochloric acid solution. The additives were added to preserve protein moieties.

2.2.4.Preparation of Breast Tumor Tissue Homogenates

The frozen tissue was washed with ice cold normal saline and then weighed, the sample was put on a Petri dish standing on ice bath, sliced and minced with scalpel and then homogenized with tris buffer solution (0.05M,pH7.2) in a ratio of 1:5 (weight: volume) using manual homogenizer.

To eliminate fibers of the connective tissues, the homogenate was filtered through multi layer nylon gauze and centrifuged at 4000 r.p.m for 30 minutes at 4°C.The supernatant and pellet were considered cytosolic and nuclear fractions respectively.

2.2.5.Determination of β₂M Levels

β_2M levels were measured in sera of all patients and normal donors using radioimmunoassay kit.

Reagents

The following reagents were equipped with the kit:

- **Tracer:** 1 vial (55ml) of ^{125}I-β_2M with 148 KBq radioactivity.
- **Standards:** 5 vials (0.5 ml each) contain from 0 to 30 mg/L of β_2M in liquid form.
- **Control:** 1 vial contains β_2M lyophilized in serum.
- **Coated tubes:** 100 tubes coated with anti-β_2M monoclonal antibodies.

Procedure

The following steps were performed to generate standard curve:

- Fifty microliters of each standard were added to the coated tubes.
- Five hundred microleters of tracer were added to each tube.
- Another tube contains 500µl of tracer was prepared separately for the determination of the total counts (T).
- All tubes were incubated for 90 minutes at 25°C with shaking at 400 r.p.m.
- After incubation, the contents of the tubes except the tube (T) were aspirated.
- The radioactivity of all tubes was measured using gamma counter.

Calculations

- The mean of the counts or B/T% values were calculated by dividing the number of bound cpm (B) of each concentration by the total counts (T).

$$\frac{B}{T} = \frac{\text{Sample Counts (B)}}{\text{Total Counts (T)}} \times 100$$

27

- The standard curve was generated by plotting the B/T% values against β_2M concentration on a semi-logarithmic paper.

2.2.6. Total Protein Determination

The total protein content of both cytosolic and nuclear fraction of breast tissue homogenates was determined by the method of Lowry etal [118] using bovine serum albumin as standard protein. The standard curve of protein concentration was constructed by measuring the absorbance of the standards at 750 nm. The straight-line equation for the standard curve was found and used for determination of protein content.

2.2.7. Preliminary Test of ^{125}I-β_2M Binding to its Receptors in Breast Tumor Homogenates

The selection of a homogenate from the collected cases for binding studies in all studied groups was done by choosing the homogenate of larger tumor mass, higher protein concentration, and higher preliminary binding reaction. So single case carrying these properties was selected from each group and it was used through the present work. Through this study, the selected tissue homogenates from malignant post-menopausal breast tumor, malignant pre-menopausal breast tumor and benign breast tumor were denoted as ThI, ThII and ThIII respectively.

Reagents

1. Tris buffer solution 0.05M, was prepared by dissolving 0.6075gm of tris(hydroxymethyl) aminomethane and 1gm of BSA in 100ml deionized water, then the pH was adjusted to 7.2 .This buffer was freshly prepared in the day of use.

2. PEG 6000 solution was prepared by dissolving 1gm in 10ml of tris buffer pH 7.2.

Procedure

1. Backgrounds of three dry tubes were measured using gamma counter.

2. Fifty micro liters (9.83×10^{-5} M) of tracer (^{125}I-β_2M) were added to the first tube that was denoted as total (T).

3. Fifty micro liters of tracer were mixed with one hundred micro liters (200µg protein) of homogenate in the second tube that was denoted as with precipitating agent (+PEG).

4. Fifty micro liters of tracer were mixed with one hundred micro liters (200 µg protein) of homogenate in the third tube that was denoted as without precipitating agent (-PEG).

5. The volume of all tubes was made up to 500 µl with tris buffer pH 7.2

6. The assay tubes were incubated for 90 minutes at 25°C.

7. Five hundred micro liters of 10% PEG6000 were added to the tube total (T) and the tube (+PEG), the two tubes were incubated for further 30 minutes.

8. All tubes were centrifuged at 4000 r.p.m for 30 minutes at 4°C.

9. The supernatant was decanted and the rim at each tube was swapped with cotton.

10. The radioactivity of the precipitated complex in each tube was counted using gamma counter.

11. Non-specific binding was accounted for by preparing the same incubation mixture with addition of 10-fold excess of unlabeled β_2M. The final volume was made up to 500µl; the assay tube was incubated for 90 minutes at 25°C.then the steps 8-10 were repeated.

12. The pellet fractions of the breast tissue homogenates were dissolved in tris-buffer solution pH 7.2 in a ratio of 1:3, the total protein content was measured according to Lowry method, and then the same steps mentioned above were followed to measure the radioactivity.

Calculations

1. The counted radioactivity (expressed as cpm) of each tube before addition of any material represents backgrounds of tubes.

2. The counted radioactivity of the tubes containing nothing but tracer (^{125}I-β_2M) represents the total counts (T).

3. The counted radioactivity of all tubes after centrifugation and drying represents the counts of total binding (TB).

4. The counted radioactivity of each tube containing tracer and an excess of unlabeled β_2M represents the counts of non-specific binding (NSB).

5. The specific binding (SB), expressed as cpm, was calculated by subtracting the counts of non specific binding from those of total binding:

$$SB \text{ (cpm)} = TB \text{ (cpm)} - NSB \text{ (cpm)}$$

6. The percentage of specific binding (SB%) was calculated as follows:

$$SB\% = \frac{SB \text{ Counts}}{\text{Total Counts (T)}} \times 100$$

2.3. Factors Affecting $^{125}I\text{-}\beta_2M$ Binding to its Receptors in Breast Tumor Homogenates

2.3.1. Effect of Different Amounts of Protein in Breast Tumor Homogenates on the Binding

Reagents

Tris buffer solution 0.05M, pH 7.2 was prepared as mentioned in section (2.2.7).

Procedure

1. Fifty micro liters of tracer $(9.83 \times 10^{-5}M)$ were counted for total radioactivity

2. One hundred micro liters of increasing amounts (50, 100,200,300,400,500 µg) of protein from the supernatant fraction of the tumor tissue homogenates were added to the assay tubes; the final volume was made up to 500 µl with tris buffer solution pH 7.2.

3. All tubes were incubated for 90 minutes at 25°C.Then the steps 8-11 mentioned in section (2.2.7) were repeated

Calculations

1. The percent of specific binding (SB%) was calculated according to the formula described in calculations of section (2.2.7).

2. The values of SB% were plotted against the corresponding protein amount included in each mixture.

2.3.2. Effect of Different $^{125}I\text{-}\beta_2M$ Concentrations on the Binding to its Receptors in Breast Tumor Homogenates

Reagents

Tris buffer solution 0.05M, pH 7.2 was prepared as mentioned in section (2.2.7).

31

Procedure

1. Increasing concentrations (3.93, 5.89, 7.86, 9.83, 11.79, 13.76 and 15.72×10^{-5}M) of ^{125}I-β_2M were counted for their radioactivity.

2. The optimum amounts of protein in breast tumor homogenates (400μg for ThI and ThII and 500μg for ThIII) were added to each tube; the final volume was made up to 500 μl with tris buffer solution pH 7.2.

3. The assay tubes were incubated for 90 minutes at 25°C.Then; steps 8 to 11 described in section (2.2.7) were repeated for the all tubes.

Calculations

1. The percent of specific binding (SB%) was calculated using the formula mentioned in calculations of section (2.2.7).

2. The resulting SB% values were plotted against the corresponding molar concentrations of ^{125}I-β_2M.

2.3.3. Effect of Medium pH on ^{125}I-β_2M Binding to its Receptors in Breast Tumor Homogenates

Reagents

Seven tris buffer solutions with different pH values (6.8, 7.0, 7.2, 7.4, 7.6, 7.8 and 8.0) were prepared as described in section (2.2.7) using hydrochloric acid solution to adjust the pH values.

Procedure

1. Tracer (11.79, 13.76 and 11.79)$\times 10^{-5}$M were added to (0.8, 0.6 and 0.8 mg/ml protein) of ThI, ThII and ThIII respectively.

2. The final reaction volume in each tube was made up to 500 μl with tris-buffer solution of different pH values (6.8-8).

3. The assay tubes were incubated for 90 minutes at 25°C.Then; steps 8 to11 mentioned in section (2.2.7) were repeated.

Calculations

1. The values specific binding percentages (SB%) were calculated using the formula mentioned in section (2.2.7).
2. The resulting SB% values were plotted against the corresponding pH values.

2.3.4.Time Course of ^{125}I-β_2M Binding to its Receptors in Tumor Homogenates

Reagents

Tris buffer solution 0.05M was prepared as mentioned in section (2.2.7). The pH was adjusted to the optimum value for each group.

Procedure

1. Tracer (11.79, 13.76 and 11.79)×10^{-5}M were added to (0.8,0.6 and 0.8 mg/ml protein) of ThI,ThII and ThIII respectively.
2. The final reaction volume in each tube was made up to 500 μl with tris-buffer solution of the optimum pH value (7.4,7.4 and 7.6) for ThI,ThII and ThIII respectively.
3. All tubes were incubated at 25°C for different time intervals (30,60,90,120,150and 180) minutes.
4. After incubation, steps 8-11 mentioned in section (2.2.7) were repeated.
5. To determine the time course of ^{125}I-β_2M binding to its receptors at different temperatures, the above steps were repeated at different temperatures (5,37 and 45) °C.

Calculations

1. The values of SB% were calculated for each tube as described in the calculations of section (2.2.7).

2. The SB% values were plotted against the time of incubation at different temperatures.

2.3.5. Effect of Different Halide Ions on ^{125}I-β_2M Binding to its Receptors in Breast Tumor Homogenates

Reagents

1. Tris buffer solutions were prepared as described in section (2.2.7).

2. Halide reagents were prepared at concentration of 0.3 M at the optimum pH for each group.

Procedure

1. Tracer (11.79, 13.76 and 11.79)$\times 10^{-5}$M were added to (0.8, 0.6 and 0.8 mg/ml protein) of ThI, ThII and ThIII respectively.

2. Fifty micro liters of 0.3 M solution of NaF were added to the reaction mixture, the final volume was made up to 500 µl with tris-buffer. The same steps were repeated for other halides solutions.

3. Steps 8-11 previously mentioned in section (2.2.7) were repeated.

Calculations

1. The values of SB% were calculated as mentioned in section (2.2.7).

2. The SB% values were plotted against the halide type.

2.3.6. Effect of Transition Metals on ^{125}I-β_2M Binding to its Receptors in Breast Tumor Homogenates

Reagents

1. Tris buffer solutions were prepared as described in section (2.2.7).

2. Solutions of $MnCl_2$, $NiCl_2$, $CuCl_2$ and $ZnCl_2$ were prepared at a concentration of 0.25 M.

Procedure

1. The reaction mixtures were prepared as described in section (2.3.5).

2. Fifty micro liters of 0.25 M of $MnCl_2$ solution were added to the reaction mixture.

3. The final volume was made up to 500 µl with tris-buffer solutions of optimum pH values for each group. The same steps were repeated for the other transition metals (i.e. $NiCl_2$, $CuCl_2$ and $ZnCl_2$).

4. Steps 8-11 previously mentioned in section (2.2.7) were repeated.

Calculations

1. The SB% values were calculated according to section (2.2.7).

2. SB% values were plotted against the transition metal type.

Results and Discussion

2.4. Determination of ^{125}I-β_2M Levels in Sera of Breast Cancer Patients

Radioimmuno assay technique was used to measure β_2M concentrations in serum samples from three groups of patients with malignant and benign breast tumors. Table (2.4) summarizes the mean levels $\pm SD$ of β_2M for the three groups compared to the control group with their (P) values.

Table (2.4). Serum β_2M levels in the three groups of patients. All details are described in section (2.2.5).

Group	Patients	No.	β_2M mg/L $\pm SD$	P values
I	Post-menopausal malignant breast tumor	28	3.82±0.94	< 0.001
II	Pre-menopausal malignant breast tumor	22	3.5±1.12	< 0.001
III	Benign breast tumor	25	3.32±0.63	< 0.001
IV	Healthy individuals	20	1.27±0.49	-

The results obtained in this study reveal that there were significant elevations (P<0.001) in β_2M concentrations in three groups of patients when compared to the control group.

Teasdale et al. [119] have shown that serum levels of β_2M significantly increase in breast cancer patients compared with normal controls. Papaioannou et al. [120] have studied patients with breast cancer in various stages and found a significant rise in β_2M serum levels in stage IV. Moreover, some authors have included β_2M measurement as a predictor of relapse in breast cancer along with CEA and ferritin [121].

However, according to Adami et al. [117], no difference in serum β_2M has been detected compared with age-matched controls.

Similarly, Kimber et al.[122] have reported no correlation between β_2M serum levels and stage of disease.

Our results agree with those of Teasdale etal.[119] and Papioannou etal.[120] and disagree with the others [117, 122]. The difference in the results may be partly accounted for by the different methods used to measure β_2M.

It has been postulated for many years that tumors can release material into the environment that compromises the immune system. Several studies have documented the shedding of various molecules from tumor cells both *in vitro* and *in vivo*.

The raised concentrations of β_2M can be attributed to its shedding into the serum due to the accelerated membrane turnover and/or cell division displayed by cancer cells.

The levels of β_2M showed only insignificant differences in the three groups of breast cancer patients when compared to each other. An explanation may lie in the possible relatedness of β_2M serum levels to tumor stage [120]. The selected patients may be not in a stage that permits β_2M leak in to their sera. The combined use of β_2M with other markers may provide high degree of marker positivity.

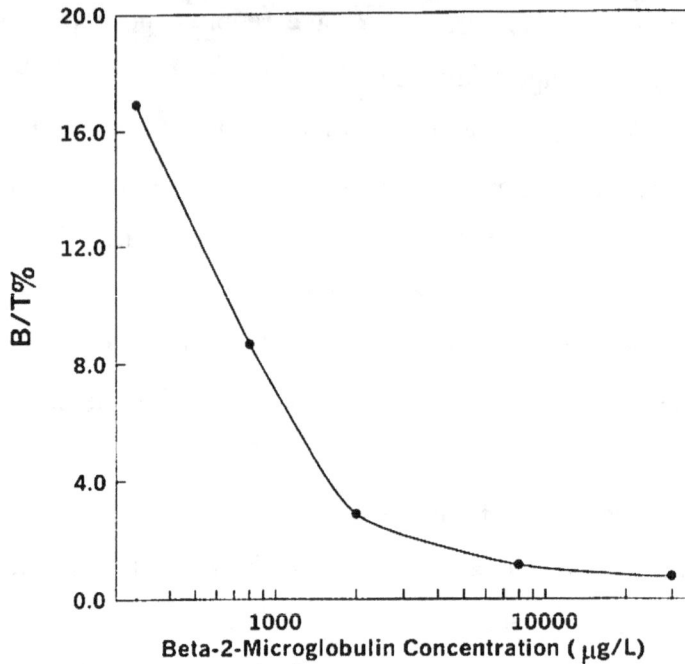

Figure (2.1). The calibration curve for determination of serum β_2M levels.

2.5. Preliminary Test of ^{125}I-β_2M Binding to its Receptors in Breast Tumor Homogenates

The presence of β_2M receptors in benign and malignant breast tumors homogenates was investigated employing the radioimmuno assay technique with our modification for tissue in which ^{125}I-β_2M acts as a ligand.

Preliminary experiments underline the ability of β_2M to bind to receptor from human tumor tissue. The binding experiments were achieved using the filtrate and the pellet fractions from each group included in this study. The results' of the preliminary binding are demonstrated in table (2.5), which reveals that β_2M receptors exist in a higher concentration in the filtrate fraction than in the pellet fraction.

The effect of PEG 6000 as a precipitating agent was tested in an attempt to aid in precipitating the formed complex, hence increasing the percent of specific binding. The results showed that PEG 6000 has no effect; accordingly the subsequent experiments were carried out using the filtrate fraction of the homogenates in the absence of PEG 6000.

Table (2.5). Results of preliminary test for ^{125}I-$\beta_2 M$ binding to its receptors in cytosolic and pellet fractions of the tumor homogenates. All details are described in section (2.2.7).

Group	SB%					
	Filtrate			Pellet		
	T	+PEG	-PEG	T	+PEG	-PEG
ThI	0.8	3.1	3.6	0.7	1.0	0.8
ThII	0.7	2.8	3.0	0.8	0.8	0.7
ThIII	0.8	3.1	3.05	0.9	0.7	0.7

T = ^{125}I-$\beta_2 M$ alone (Tracer).
+PEG= presence of precipitating agent (PEG6000).
-PEG = absence of precipitating agent.

2.6. Factors Affecting ^{125}I-$\beta_2 M$ Binding to its Receptors in Breast Tumor Homogenates

2.6.1. Effect of Various Concentrations of Protein in Breast Tumor Homogenates

These experiments were carried out to examine the effect of different protein concentrations on the binding at fixed concentration of ^{125}I-$\beta_2 M$ and to select the optimum one which gives the highest SB% value.

Increasing protein concentrations from three types of homogenates (i.e. ThI,ThII and ThIII) exhibited the same behavior revealed in figure (2.2).There was an increase in the specifically bound ^{125}I-$\beta_2 M$ expressed as (SB%) with increasing protein concentrations until reaching the point of maximum binding where no more protein can be bound to the labeled ligand.

The hyperbolic curves shown in this study are generally similar to those obtained from other studies dealing with biomolecular interactions [123].

The optimum protein concentrations were determined and used in the subsequent experiments.

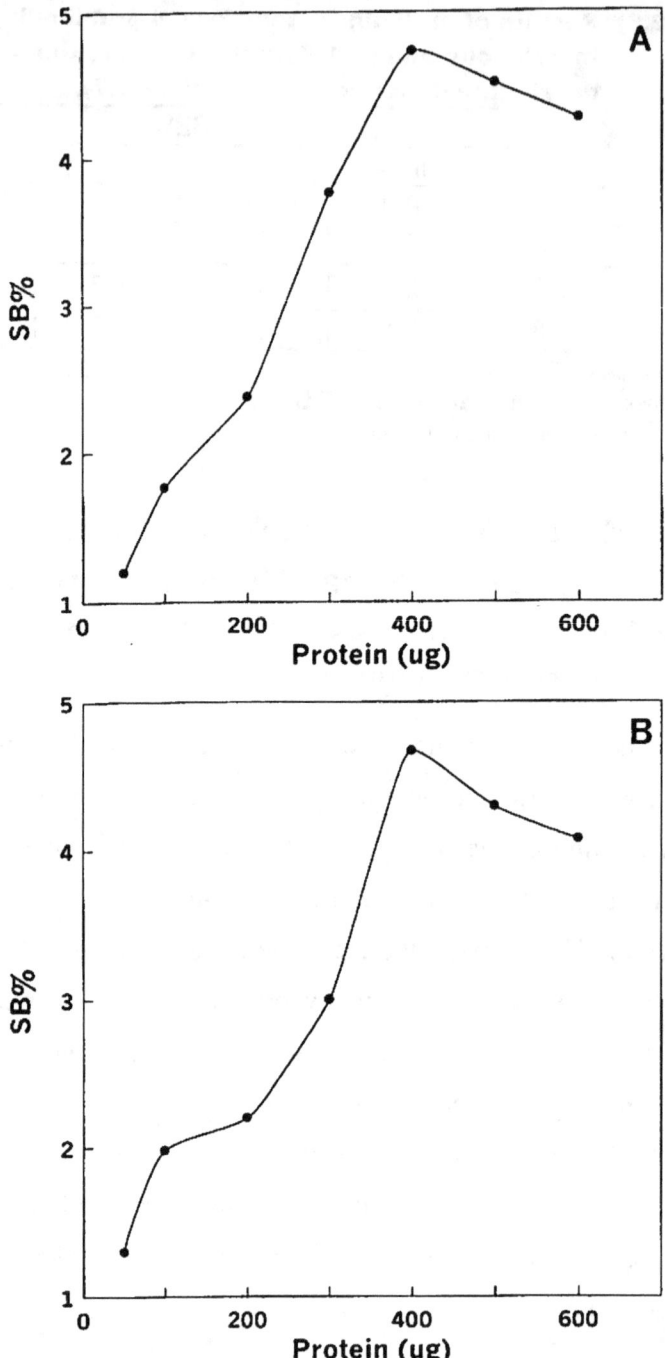

Figure (2.2). The effect of protein concentration on ^{125}I-β_2M binding to its receptors in breast tumor homogenates, A) ThI, B) ThII and C) ThIII. All details are described in section (2.3.1).

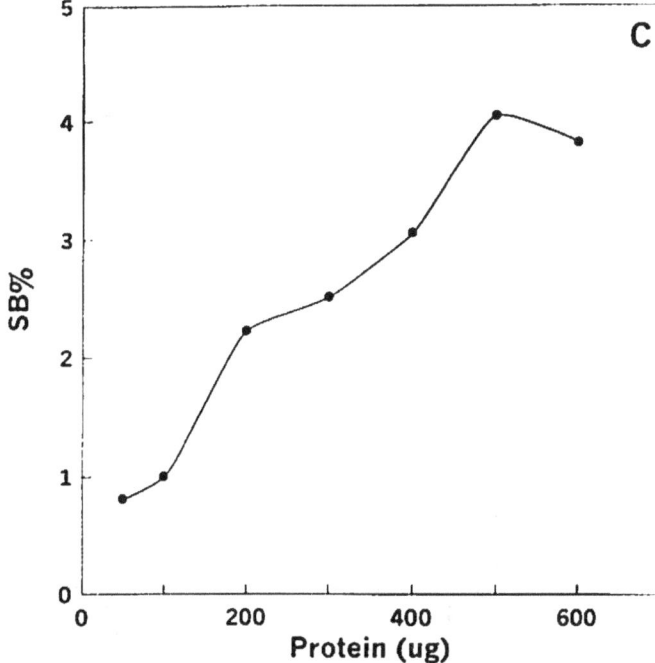

Figure (2.2). Continued.

2.6.2. Effect of Different ^{125}I-β_2M Concentrations on the Binding to its Receptors in Breast Tumor Homogenates

This set of experiments was carried out by incubating the previously-determined optimum concentrations of protein with increasing ^{125}I-β_2M concentrations.

Figure (2.3) reveals the relationship between the increasing ^{125}I-β_2M concentrations and the corresponding values of SB% in three types of breast tumors homogenates. A possible explanation for these data takes into account Hook effect, according to which a biphasic response curve might result from multivalent biomolecular interactions[124]. β_2M may presumably act as a multivalent ligand considering the studies which have focused on the binding characteristics of β_2M[125].

The optimum tracer concentration was 11.79×10^{-5} M for ThI and ThIII, while it was 13.76×10^{-5} M for ThII. Accordingly, these concentrations were used in all subsequent experiments.

41

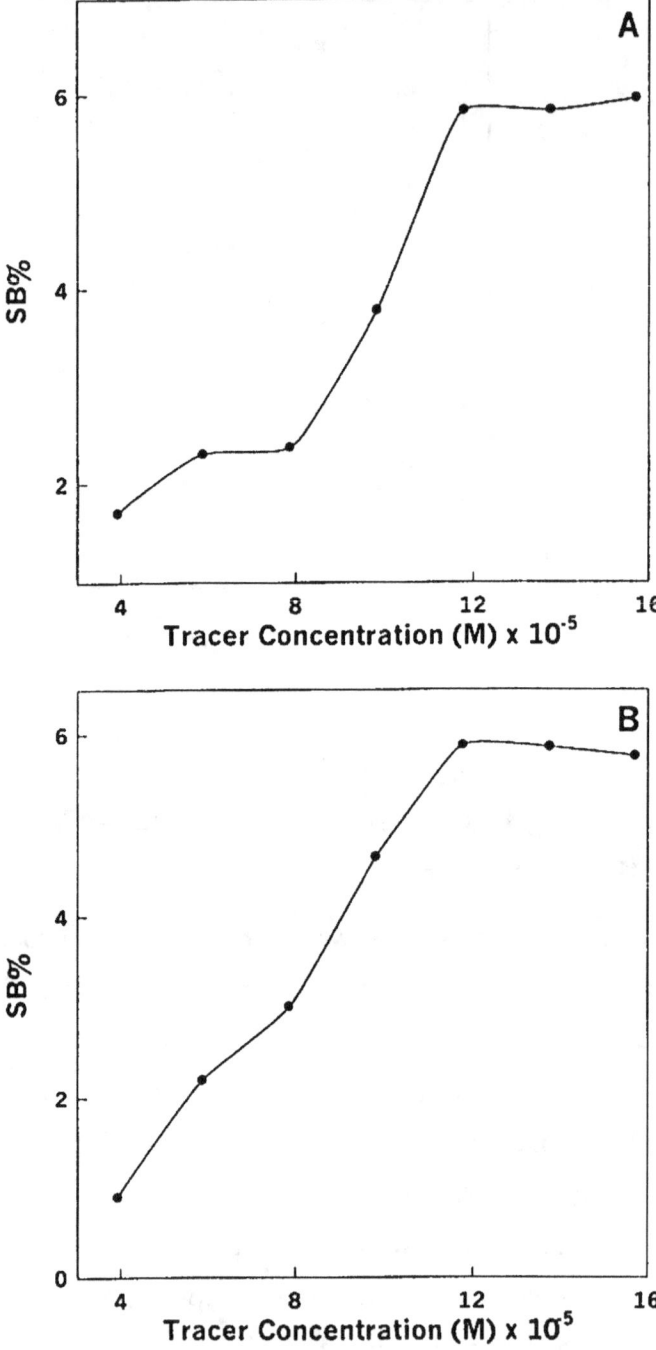

Figure (2.3). The effect of ^{125}I-β_2M concentration on the binding to its receptors in breast tumor homogenates, A) ThI, B) ThII and C) ThIII. All details are described in section (2.3.2).

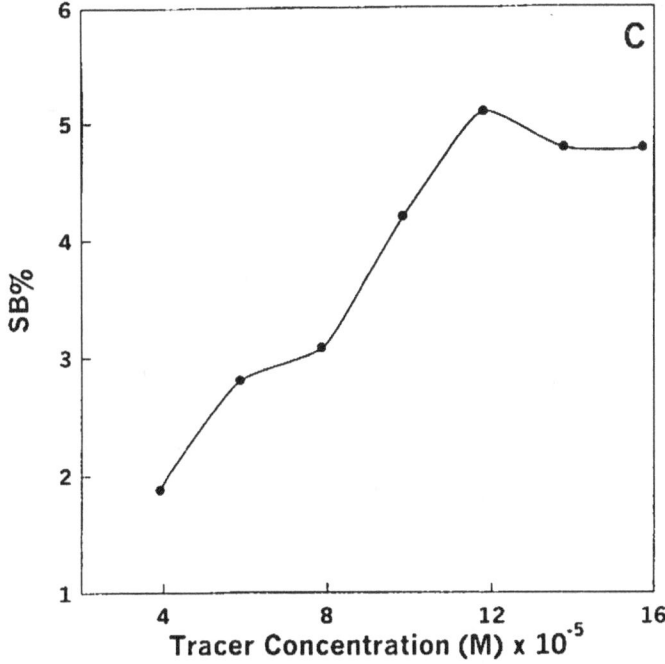

Figure (2.3). Continued.

2.6.3. Effect of pH on ^{125}I-$\beta_2 M$ Binding to its Receptors in Breast Tumor Homogenates

The effect of the medium pH on the binding was tested using tris buffer solutions with pH values ranging from 6.8 to 8. Figure (2.4) shows the typical bell-shaped curve obtained from the three types of breast tumors homogenates included in this study.

It is not surprising that pH will influence the binding reaction. The binding sites on $\beta_2 M$ molecules and its receptors are frequently composed of ionizable groups (in the amino acids side chains) that must be in the proper ionic form in order to maintain the conformation of the binding sites, consequently, changing the reaction pH will induce protonation-deprotonation process within the ionizable groups resulting in a destructed conformation.

The slight shift in the optimum pH values observed in figure (2.3) may be a consequence for a slight alteration in amino acid composition of the receptors according to their site of origin. The pH in the subsequent experiments was adjusted at the optimum values.

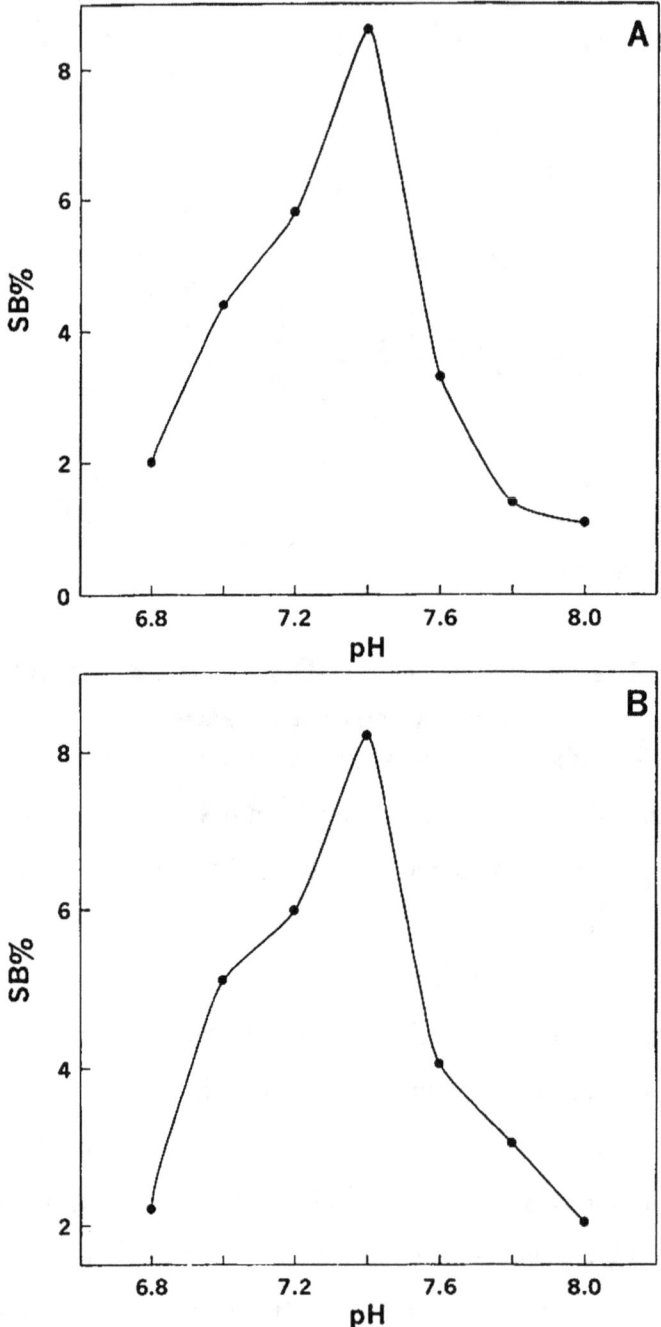

Figure (2.4). The effect of medium pH on ^{125}I-β_2M binding to its receptors in breast tumor homogenates, A) ThI, B) ThII and C) ThIII. All details are explained in section (2.3.3).

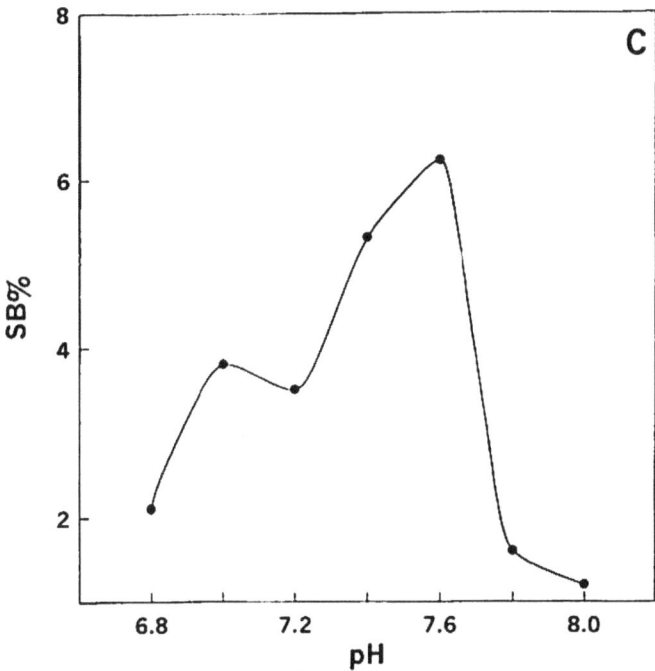

Figure (2.4). Continued.

2.6.4. Time Course of ^{125}I-β_2M Binding to its Receptors in Breast Tumor Homogenates

Time and temperature dependency of the binding of ^{125}I-$\beta 2$M to its receptors was investigated. The results depicted in figure (2.5) reveal that the binding is a time and temperature dependent process with a maximum binding occurs at 37°C and incubation time reaches 90 minutes for all types of breast tumor homogenates.

Most chemical reactions proceed at a faster velocity (i.e. give higher concentration of product) as the temperature is raised. An increase in temperature imparts more kinetic energy to the reactant molecules resulting in more productive collisions per unit time. β_2M and its receptor behaved similarly, up to a point. Prolonged incubation time (i.e. above 2 hours) and higher temperature (i.e. above 37°C) seem to influence the binding process, presumably, through their destructive effect on different intramolecular forces within β_2M and the receptors, hence lack of the regular three-dimensional structure and/or their effect on the non-covalent forces that hold the two proteins together resulting in irreversible dissociation of the complex.

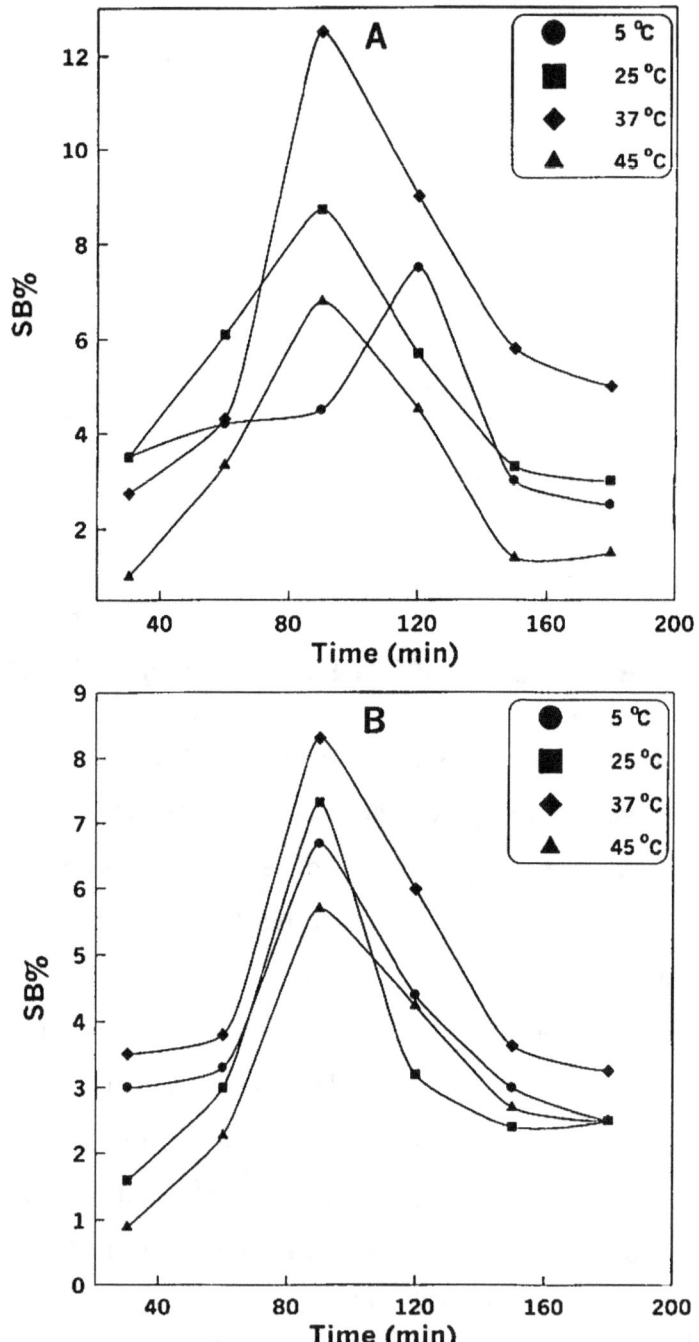

Figure (2.5). The time course of ^{125}I-β_2M binding to its receptors in breast tumor homogenates, A) ThI, B) ThII and C) ThIII. All details are described in section (2.3.4).

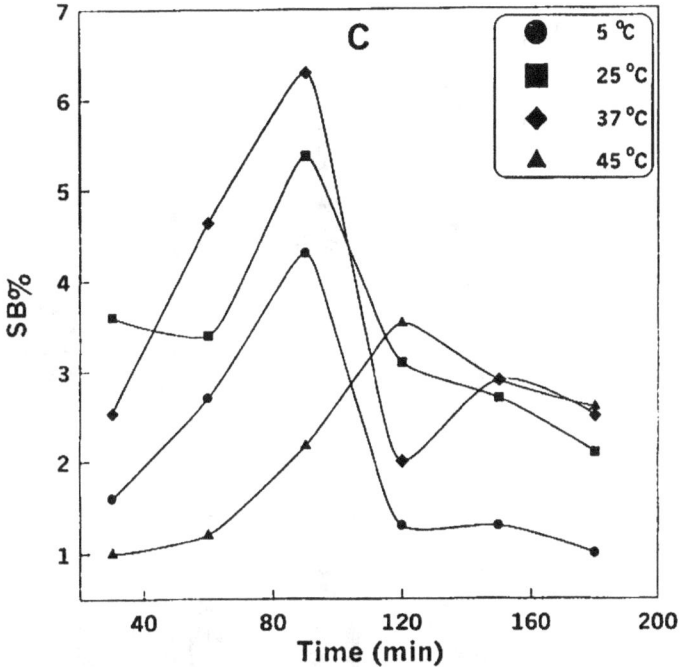

Figure (2.5). Continued.

2.6.5. Effect of Halides Ions of ^{125}I-β_2M Binding to its Receptors in Breast Tumors Homogenates

The effect of sodium halides on the binding was examined. The results obtained in these experiments, figure (2.6) indicate that halides ions diminish the binding activity of ^{125}I-β_2M and the receptors according to the sequence:

$$I^- > Br^- > Cl^- > F^-$$

The most likely explanation for these results is that halides ions may exist in the interface between the radiolabeled β_2M and its receptors resulting in a subtle distortion in the regular three dimensional structure making it difficult for the two proteins to bind. The distortive effect of choatropic anions has been previously reported [126].

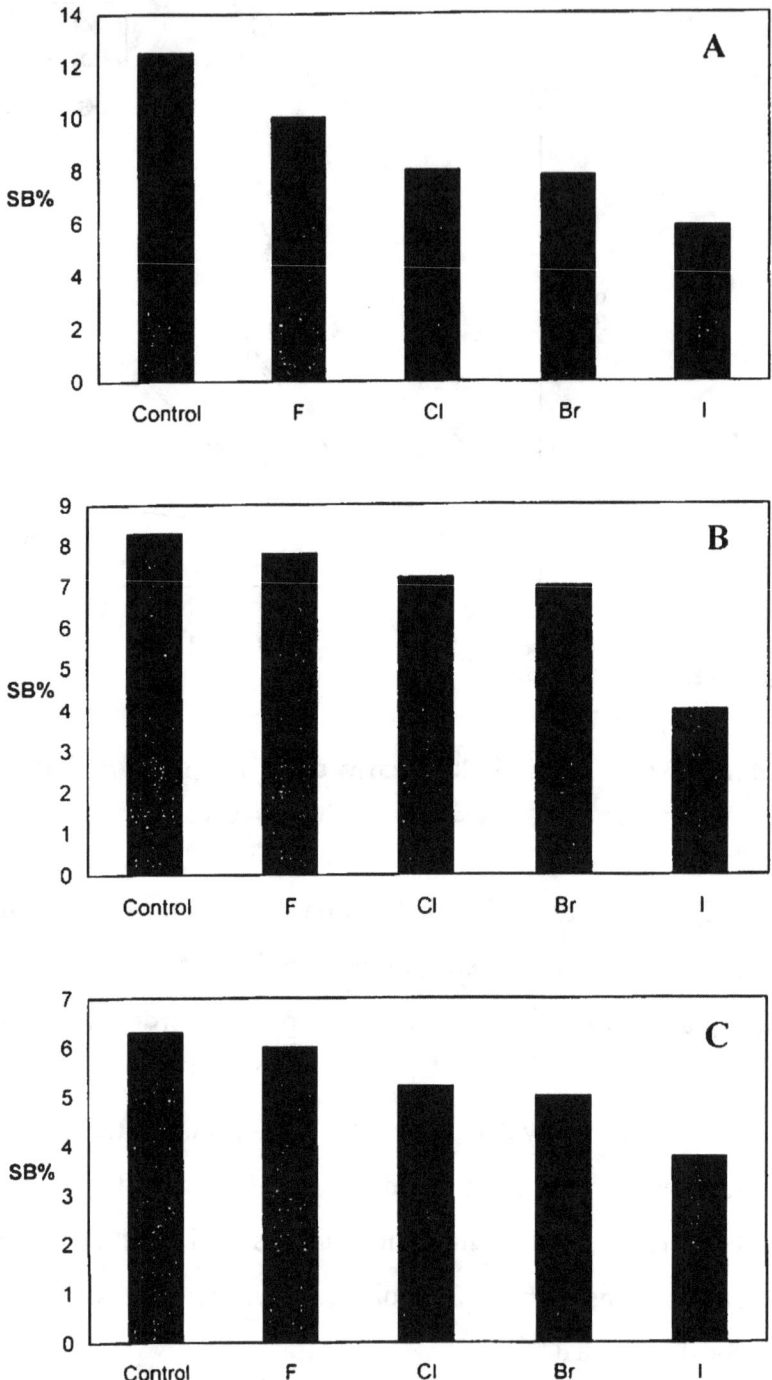

Figure (2.6). The effect of halides ions on ^{125}I-β_2M binding to its receptors in breast tumor homogenates, A) ThI, B) ThII and C) ThIII. All details are described in section (2.3.5).

2.6.6. Effect of Transition Metals Ions on ^{125}I-β_2M Binding to its Receptors in Breast Tumors Homogenates

Figure (2.7) shows the effect of some transition metals ions on the binding of ^{125}I-β_2M to its receptors in breast tumor homogenates. It is known that transition metals ions can exploit their d-orbital to coordinate with the donor atoms of different ligands. Proteins display more than one atom available for this coordination.

According to this fact, one could envisage the formation of complexes in which the metal ion coordinates to both ^{125}I-β_2M and the receptors in such complex the two protein molecules share the same central ion resulting in higher amount of the detectable complex. Another explanation suggests that the formation of coordinating complex might bring the two proteins closer and play a role in proper orientation of the binding sites on reactants molecules (^{125}I-β_2M and the receptors)

The effect of transition metals ions was in accordance with Irving Williams's series of stability [127], so Cu^{+2} ions have the strongest effect on binding among other ions, hence the highest SB% value.

$$Mn^{2+} < Ni^{2+} < Cu^{2+} > Zn^{2+}$$

49

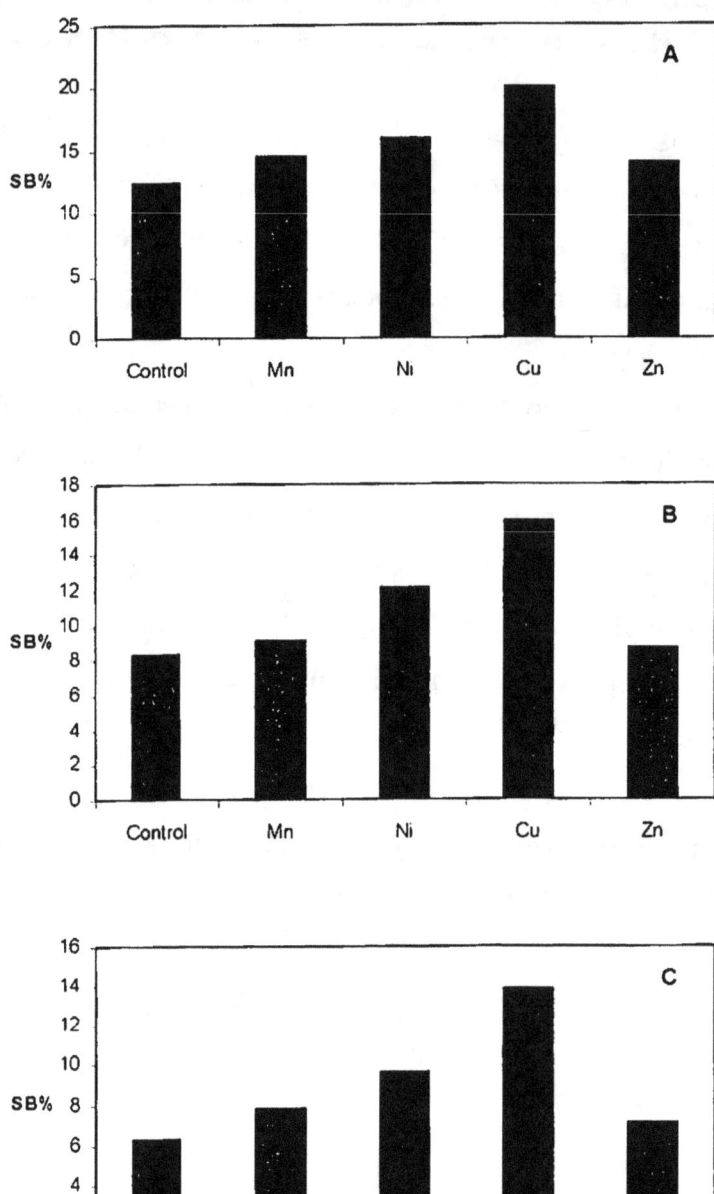

Figure (2.7). The effect of transition metals ions on ^{125}I-β_2M binding to its receptors in breast tumor homogenates, A) ThI, B) ThII and C) ThIII. All details are described in section (2.3.6).

Isolation of ^{125}I-β_2M Receptors by Gel Filtration Chromatography

Chapter Three

Isolation of β₂M Receptors by Gel Filtration Chromatography

Introduction

Column chromatography represents an extremely efficient family of techniques for the isolation and identification of proteins in biological extracts where they exploit the differences in physical and chemical properties [128].

The purification of proteins is the most common use of gel filtration chromatography due to the ability of a gel to fractionate molecules on the basis of size and shape [129].

The efficiency of the isolation method is given by the purification factor by which the specific activity of the preparation has increased [130].

Unfortunately, no similar studies are available dealing with the receptors for β₂M. However, gel filtration chromatography has been employed to isolate the β₂M receptor from murine cells. The isolated receptor has been found to comprise a 48 KD glycoprotein. Occasionally, variable quantities of non-glycosylated 25 KD component was also present [106].

The scope of this chapter is to retain as much of the β₂M receptors as possible while getting rid of as much of the other proteins as possible and to definitely establish the optimum conditions of their binding reaction.

Materials and Methods

3.1. Materials

3.1.1. Chemicals

The following chemicals and reagents were used in the experiments of this chapter.

Table(3.1). Chemicals used.

Chemical	% Purity	Company
β_2M RIA kit	-	Immunotech.(Czech Rep.)
BSA	99	Fluka (Switzerland)
Na,K-tartrate	99	Fluka (Switzerland)
NaOH,Hcl and NaN$_3$	98	BDH (U.K.)
Tris(hydroxymethyl)aminomethane	99	BDH (U.K.)
Folin-ciocalteau	99	E.Merck Dastmstopt
Low molecular weight determination kit	-	Pharmacia (Sweeden)

3.1.2. Instruments

All instruments mentioned in section (2.1.2) were also used in the experiments of this chapter.

3.1.3. Tissue Homogenate

A breast tumor homogenate from a post-menopausal patient with infiltrative ductal carcinoma was used in the experiments of this chapter.

3.2. Methods

3.2.1.Preparation of the Column

A sephadex G-100 column was used to isolate $\beta_2 M$ receptors, the dimensions of the column were chosen according to the following equation:

$$\text{Diameter} = \sqrt{\frac{m}{10}} \quad \text{and} \quad L = 30 \times \text{diameter}$$

Where:

m: amount of protein in milligrams.

L: length of the column.

3.2.2. Preparation of the Eluting Buffer

Tris-HCl buffer solution (0.05M, pH 7.4) containing 5mM EDTA and 0.02% sodium azide as an antibacterial agent was prepared by dissolving 0.6075 gm of tris (hydroxymethyl) amino methane, 0.181 gm of EDTA, and 0.02% (w:v) sodium azide in 100 ml distilled water, the pH was adjusted to 7.4 using concentrated hydrochloric acid solution.

3.2.3.Preparation of the Gel

The gel was prepared by allowing the pre-swollen gel to swell again in tris buffer solution (0.05 M) pH 7.4, then it was left to settle and the excess of the buffer was decanted. This step was repeated several times. The gel was degassed using evacuating pump and slurry was left for 24 hours to be equilibrate with buffer.

The swollen gel was suspended and carefully poured into a vertical glass column (0.9 × 30 cm.) down the wall using a glass rod. After the gel has settled, the column was equilibrated with tris buffer for 24 hours.

3.2.4. Void Volume Determination

The void volume of the column was determined by using blue dextran 2000 at a concentration of 2 mg/ml dissolved in tris-buffer pH7.4, then the elution was carried out with the same buffer at a flow rate of 10ml/hr. Fractions of 1ml were collected and their absorbencies were measured at 600 nm.

3.3. Gel Filtration Chromatographic Studies

3.3.1. Isolation of β₂M Receptors by Sephadex G-100 Column

Reagents

Tris buffer solution pH 7.4 containing 0.02% Na_3N was prepared as previously described in section (3.2.2.).

Procedure

1. The sample of tissue homogenate (555µl) containing approximately 10 mg protein was applied to the surface of the gel, and then equilibrated with 0.05M tris buffer pH 7.4.

2. The sample was eluted by using the same tris buffer with a flow rate of 10 ml /hr. Fractions of 1ml volume were collected, and the gel filtration was carried out at 10°C.

3. The protein content of each fraction was determined according to Lowry method; the absorbance of each fraction was measured at 280 nm.

4. The assay method was carried out using the collected fractions in order to identify the one that contains the β₂M receptors.

Calculations

1. The percentage of specific binding (SB%) was calculated for each fraction as mentioned in section (2.2.7).

2. The SB% values were plotted against the fraction number.

3. The purification fold of β_2M receptors was calculated by the equation:

$$\text{Purification Fold} = \frac{\text{Bound of purified receptors/mg Protein}}{\text{Bound of crude receptors/mg Protein}} \times 100$$

3.3.2. Isolation of (^{125}I-β_2M / Receptor) Complex

Reagents

Tris buffer solution (0.05M) pH 7.4 containing 0.02% Na_3N was prepared as mentioned in section (3.2.2.).

Procedure

1. ^{125}I-β_2M was reacted with its receptors in the crude sample of breast tumor homogenate at their optimum conditions.

2. At the end of the reaction, 555 ml of the reaction mixture were applied to the surface of the gel, equilibrated with tris buffer pH 7.4, a flow rate of 10 ml /hr was adjusted.

3. Fractions of 1 ml were collected. The radioactivity of each fraction was then counted.

4. Unreacted ^{125}I-β_2M was poured into the column, and the radioactivity of each fraction was counted.

Calculations

The counted radioactivity as (cpm) was plotted against the corresponding fraction number.

3.3.3. Determination of Molecular Weight for the Isolated Receptors

The molecular weight of the isolated receptors was determined using pharmacia calibration kit. This kit contains highly purified low molecular weight proteins (albumin, ovalbumin, cymotrypsinogen and ribonuclease).

Reagents

1. Tris buffer solution (0.05M) pH 7.4 containing 0.02% Na_3N was prepared as described in section (3.2.2).

2. Each calibration kit protein was dissolved to a concentration of 5mg/ml, a measured volume of tris buffer pH 7.4 was added to the appropriate preweighed protein mixture (ribonuclease and ovalbumin) and (chymotrypsinogen and albumin).

Procedure

1. At the first run, 555µl of the mixture (albumin and cymotrypsinogen) were applied to the sephadex G-100 column.

2. Fractions of 1ml volume were collected at a flow rate of 10 ml /hr.

3. At the second run, 555µl of the mixture (ribonuclease and ovalbumin) were applied to the sephadex G-100 column. Then step 2 was repeated.

Calculations

The K_{av} values for standard proteins and isolated receptors were calculated using the formula:

$$k_{av} = \frac{V_e - V_o}{V_t - V_o}$$

Where:

V_o : void volume.

V_e : elution volume of each protein.

V_t : total gel bed volume which was calculated as follows:

$$V_t = \left(\frac{d}{2}\right)^2 \times 3.14 \times h$$

d=0.9 cm.

h=30 cm.

3.4. Determination of the Optimum Conditioners for ^{125}I-β₂M Binding to the Isolated Receptors

The optimum conditions of ^{125}I-β₂M binding to its isolated receptors including protein concentration, tracer concentration, pH, temperature and time were determined using the same experiments previously mentioned in section (2.3) of chapter two.

Results and Discussion

3.5. Isolation of β_2M Receptors by Sephadex G-100 Column

The isolation of β_2M receptors was achieved using gel filtration chromatography technique in which the protein content of the malignant post-menopausal breast tumor homogenate was separated according to the differences in molecular weight. Figure (3.1) shows the elution profile of blue dextran 2000 that was used to determine the void volume of sephadex G-100 column employed in these experiments. The void volume was found to be 10 milliliters.

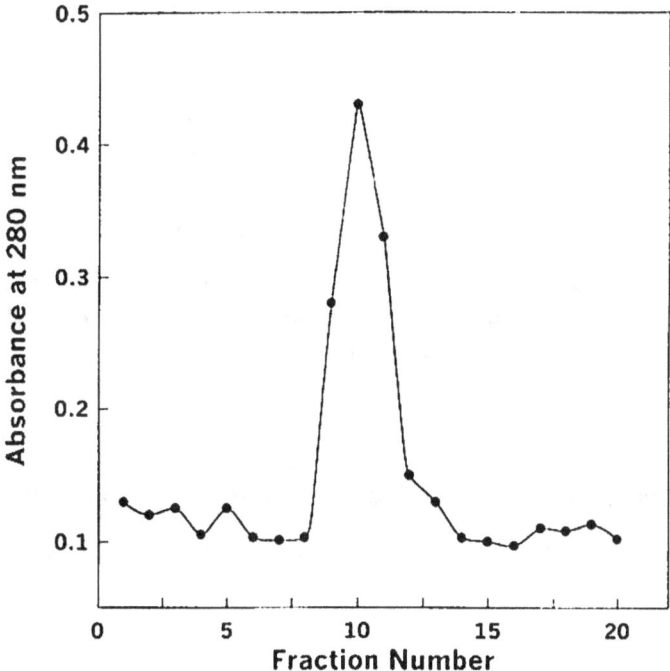

Figure (3.1). The elution profile of blue dextran, using Sephadex G-100 gel, 10ml/hr flow rate, tris-buffer pH 7.4 and at 25°C. All details are described in section (3.2.4).

The elution profile of the breast tumor homogenate, figure (3.2), revealed two main peaks at fraction number 11 and 26 when the collected fractions were measured at 280 nm. representing the protein content of the homogenate.

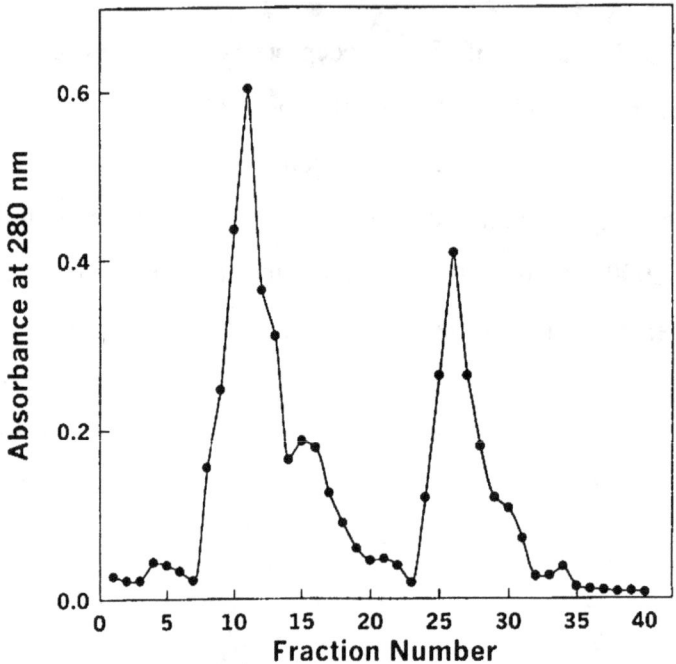

Figure (3.2). The elution profile of the breast tumor homogenate, using Sephadex G-100 gel, 10ml/hr flow rate, tris-buffer pH 7.4 and at 25°C. All details are described in section (3.3.1).

The collected fractions were reacted with ^{125}I-β₂M at their optimum conditions previously determined in section (2.3), the binding assays gave rise to figure (3.3) in which two peaks can be seen at fraction number 14 and 25 indicating the presence of β₂M receptors in those fractions.

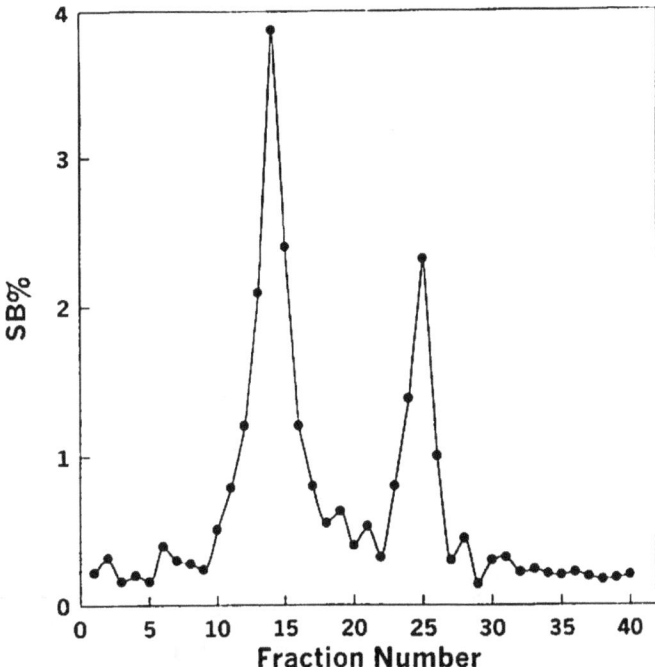

Figure (3.3). The binding of the eluted fraction of the breast tumor homogenate, using Sephadex G-100 gel, 10ml/hr flow rate, tris-buffer pH 7.4 and at 25°C. All details are described in section (3.3.1).

3.6.Isolation of (^{125}I-β_2M/Receptor) Complex

^{125}I-β_2M and its receptors in crude breast tumor homogenate were incubated at their optimum conditions; the resulting complex was isolated from the free ^{125}I-β_2M by its application to the gel filtration column. The elution profile shown in figure (3.4) reveals two peaks related to two complexes and a third peak at fraction number 36 related to the free ^{125}I-β_2M.

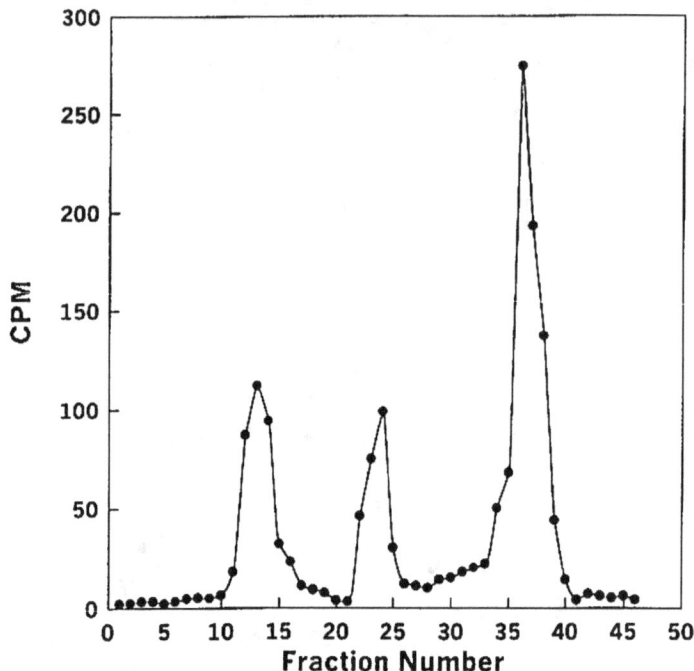

Figure (3.4). The elution profile of the complex (¹²⁵I-β₂M/receptor) prepared from the crude sample, using Sephadex G-100 column, 10ml/hr flow rate, tris-buffer pH 7.4 and at 25°C. All details are described in section (3.3.2).

The fractions forming the two peaks in figure (3.2) were pooled separately and reacted with ^{125}I-β$_2$M at the optimum conditions. Each complex resulted from the binding reaction was separated from the free ^{125}I-β$_2$M by gel filtration using the same column. Figure (3.5) shows the elution profile of the two complexes prepared in this experiment with their peaks appeared at fraction number 14 and 25.These results are in agreement with those obtained from the experiment of the isolation of β$_2$M receptors from crude sample, figure (3.4), they are both confirm the presence of two protein components that are able to be bound to β$_2$M.

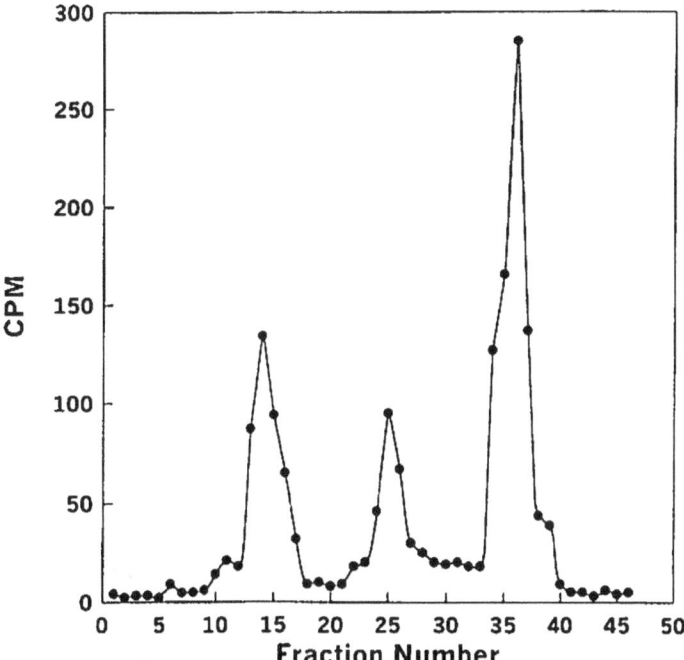

Figure (3.5). The elution profile of the complex (^{125}I-β$_2$M/receptor) prepared from the pooled peaks, using Sephadex G-100 column, 10ml/hr flow rate, tris-buffer pH 7.4 and at 25°C. All details are described in section (3.3.2).

To ascertain the position of the free tracer peak, ^{125}I-β$_2$M was applied to the gel filtration column. The results gave rise to figure (3.6) in which a single peak appeared at fraction number 37.

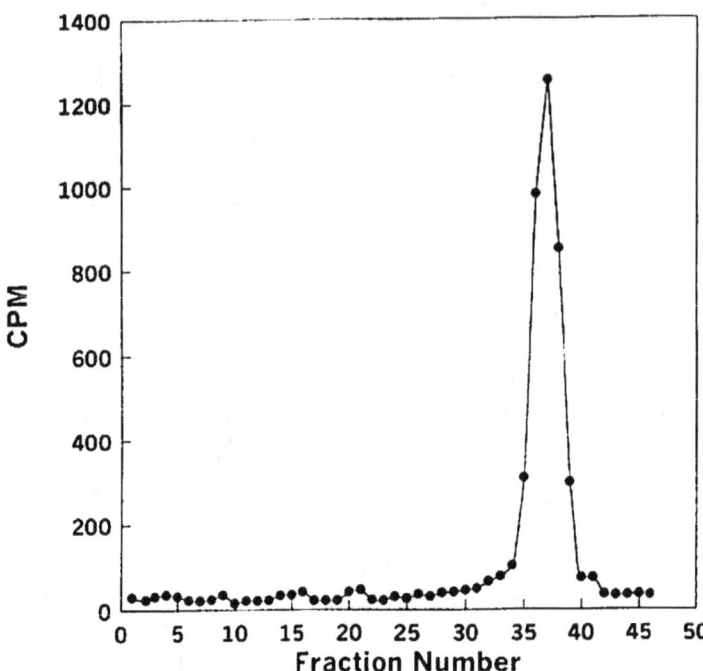

Figure (3.6). The elution profile of the free¹²⁵I-β₂M, using Sephadex G-100 column, 10ml/hr flow rate, tris-buffer pH 7.4 and at 25°C. All details are described in section (3.3.2).

Gel filtration chromatographic studies indicate clearly that ^{125}I-β_2M could bind to two protein components (receptors) having a certain difference in molecular weight. To explain further these results, it could be said that the low molecular weight component might be a degradation product of the high molecular weight one. Another possibility, the obtained results may indicate that β_2M receptors exist in two isoforms. Whatever the matter, the specific binding per milligram protein has increased, table (3.2).

It seems that the presence of the receptors with many other proteins in the crude homogenate could markedly attenuate their binding affinity towards ^{125}I-β_2M, the gel filtration process act to decrease the number of proteins in the reaction mixture and hence increase the probability of ^{125}I-β_2M to bind its receptors.

Table (3.2). Purification parameters of β_2M receptors isolated from a post-menopausal breast tumor homogenate (ThI) using gel filtration chromatography. All details are described in section (3.3.1).

Receptor	Protein (mg)	Specific binding	Specific binding /mg protein	Purification fold
ThI	0.40	12.5	31.25	1.0
Protein I	0.04	7.4	185.00	6.0
Protein II	0.04	5.6	140.00	4.5

3.7. Determination of Molecular Weight

Figure (3.7) shows the elution profile of the standard proteins used in this experiments. The K_{av} value of each protein was plotted against logarithm of molecular weight of the corresponding protein .The straight line equation generated from the resulting plot, figure (3.8), was used to determine the molecular weight of the isolated β_2M receptors. This experiment showed that the first receptor has a molecular weight of 52.8 KD while the second of 29 KD.

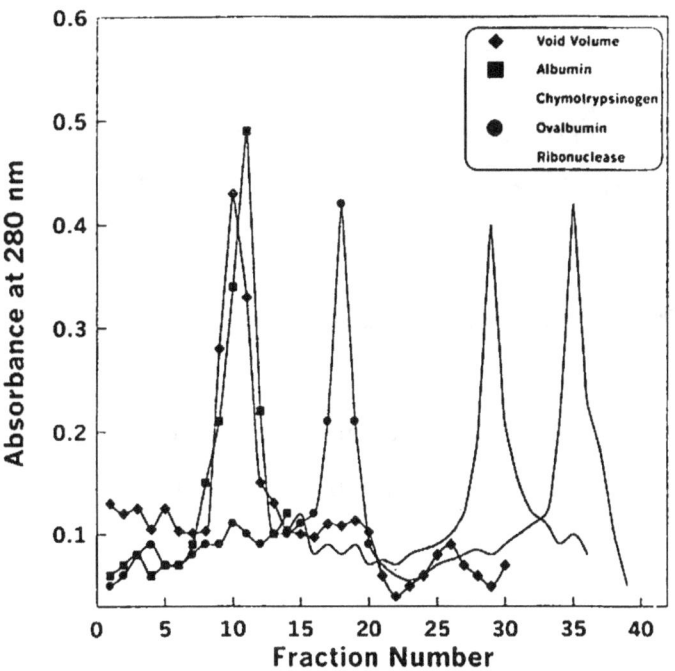

Figure (3.7). The elution profile of standard proteins used in molecular weight determination, using sephadex G-100 column, 10 ml/hr flow rate, tris-buffer pH 7.4 and at 25°C. All details are described in section (3.3.3).

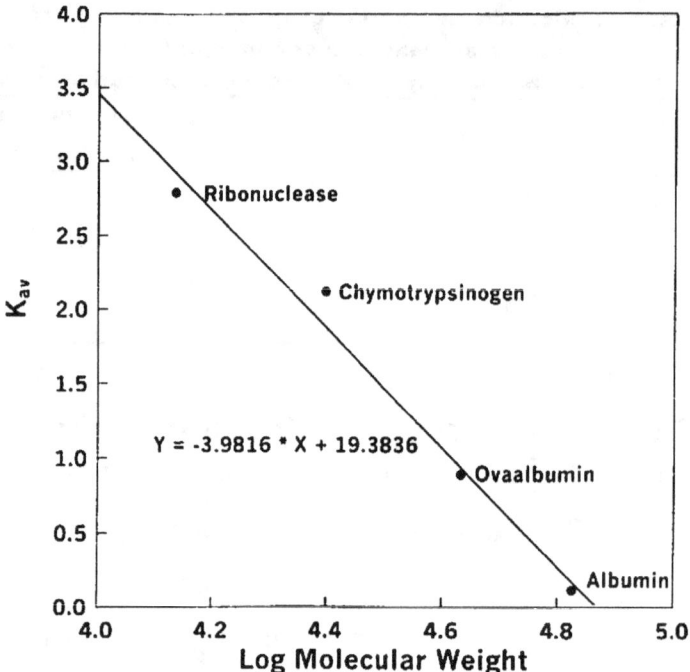

Figure (3.8). Calibration curve for molecular weight determination by gel filtration chromatography using low molecular weight Pharmacia Calibration Kit.

3.8. Determination of the Optimum Conditions for ^{125}I-β₂M Binding to the Isolated Receptors
3.8.1. Determination of Optimum Receptor Concentration

The effect of increasing amounts of isolated β₂M receptors (protein I and Π) was investigated, figure (3.9) shows an increase in the specific binding (SB%) with increasing receptor concentration (for the two proteins) until reaching the saturation region where the binding remains unchanged whatever the concentration is.

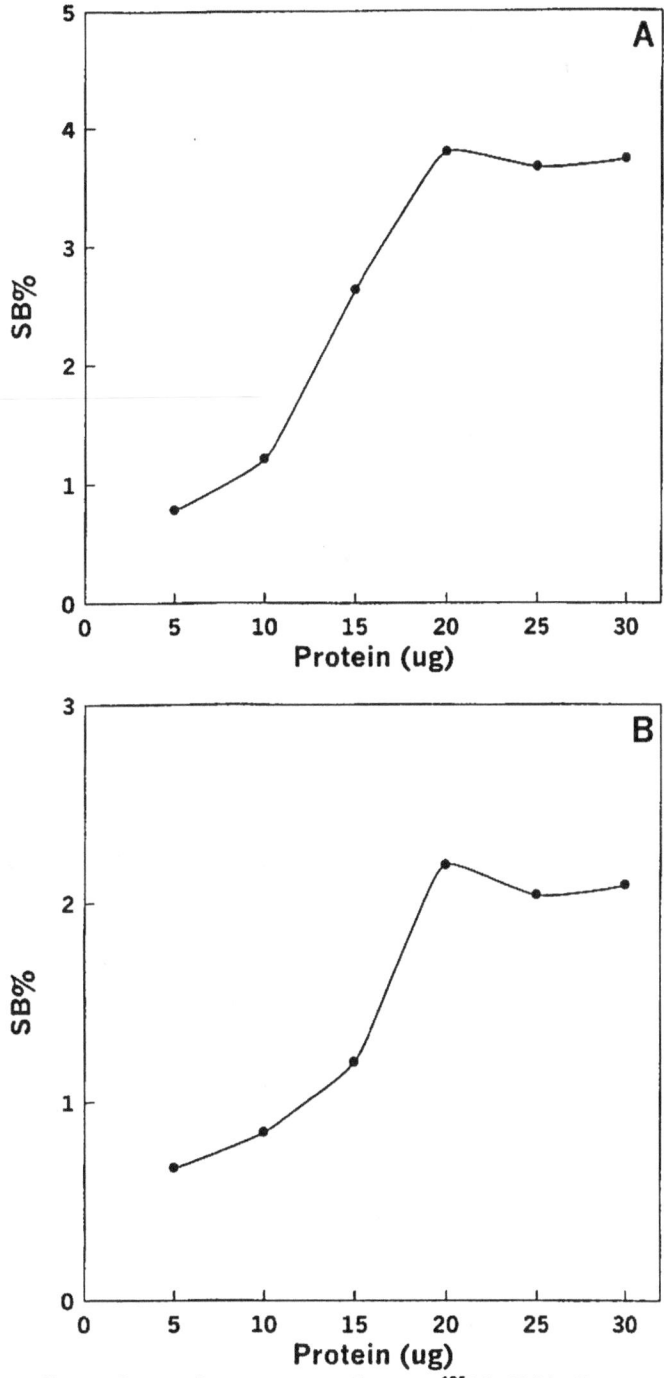

Figure (3.9). The effect of protein concentration on ^{125}I-β₂M binding to the isolated receptors, A) Protein I, B) Protein II. All details are described in section (3.4) and (2.3).

3.8.2.Determination of Optimum ^{125}I-β_2M Concentration

Different concentrations ^{125}I-β_2M were used in the reaching mixture to investigate its effect on the binding. Figure (3.10) shows a behavior similar to that obtained with the crude homogenate in which a biphasic response curve may indicate a probable multivalent character of β_2M.

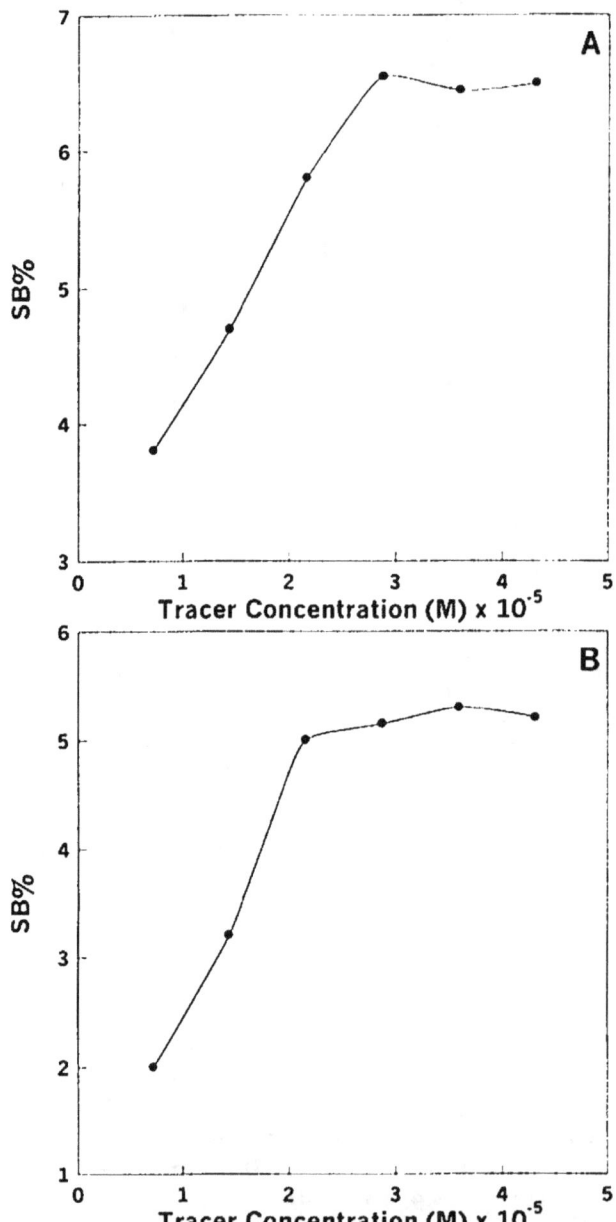

Figure (3.10). The effect of ^{125}I-β_2M concentration on the binding with the isolated receptors, A) Protein I, B) Protein II. All details are described in section (3.4) and (2.3).

3.8.3. Determination of the Optimum pH

Figure (3.11) demonstrates the effect of medium pH on ^{125}I-β_2M binding to the partially purified receptors. It is noticeable that the percentages of specific binding decline above and below the optimum value of pH 7.4. Generally, changing the environment pH can lead to a dramatic changes in protein conformation through the induction of protonation-deprotonation process within the ionizable groups resulting in formation of improper ionic forms.

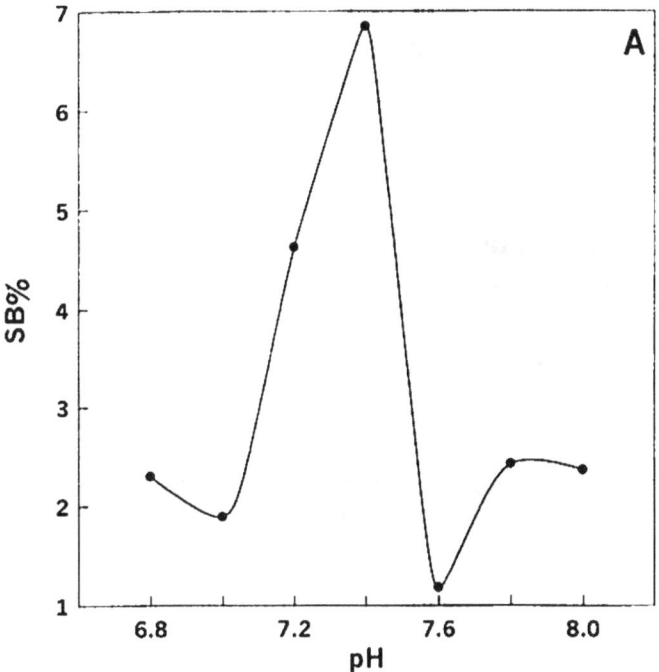

Figure (3.11). The effect of medium pH on ^{125}I-β_2M binding to the isolated receptors, A) Protein I, B) Protein II. All details are described in section (3.4) and (2.3).

69

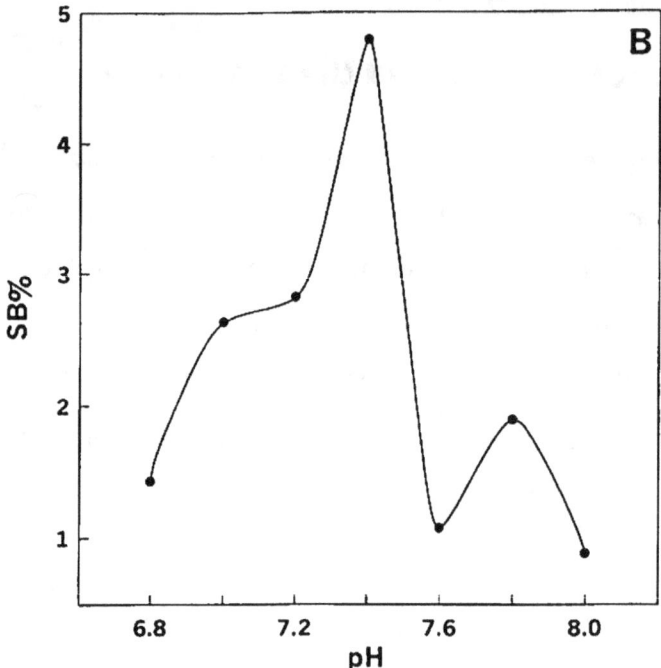

Figure (3.11). Continued.

3.8.4. Time Course of ^{125}I-β₂M Binding to its Isolated Receptors

The results of time course experiments shown in figure (3.12) indicate that the binding reaction is a time and temperature dependent process. It was found that the optimum reaction occurs at 37°C after incubation for 90 minutes. These results are in agreement with those obtained when the crude receptors were used (section (2.6.4)).

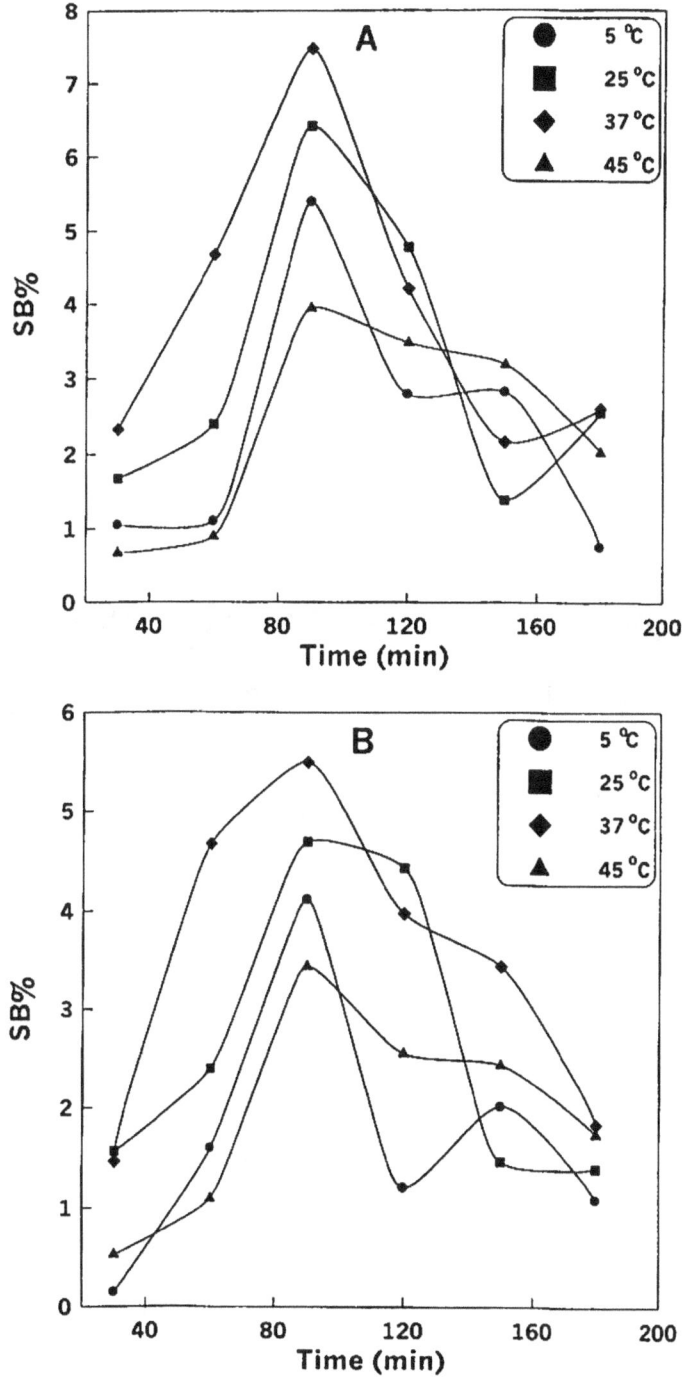

Figure (3.12). Time course of ¹²⁵I-β₂M binding to the isolated receptors, A)Protein I, B) Protein II. All details are described in section (3.4) and (2.3).

CHAPTER **4**

Kinetic & Thermodynamic Studies on the Binding of ^{125}I-β_2M to its Receptors

Chapter Four

Kinetic & Thermodynamic Studies on the Binding of ^{125}I-β_2M to its Receptors

Introduction

Many biological processes require the formation of protein-protein complexes. A variety of dynamic conformational changes are observed in proteins upon binding which involve two major interacting energy components, namely the electrostatic and the hydrophobic forces that are best understood by a study of the thermodynamic parameters of the binding [131].

Also, the measurement of the binding affinity and stoichiometry gives essential information on topics that range from mechanism to specificity [132].

Equilibrium approaches can be used to study the interaction of the protein and the ligand. The study begins by first writing an equation for the dissociation constant of the binding reaction. The equation is then restated in terms of a particularly useful parameter, and rearranged to a form that enables the data to be presented usually in a straight-line plot [133]. There are several graphical methods to linearize the binding data, the most often used is the one described by Scatchard; the ratio of specific binding to concentration of free radioliganal is plotted against the specific binding [134,135].

Determination of the thermodynamic parameters for ^{125}I-β_2M binding to its receptors under equilibrium conditions can provide information concerning the initial state of β_2M in aqueous buffer and the final state when it is bound to the receptor. Kinetic measurements supplement that information, in that they measure the differences between those states and an intermediate activated complex. Towards this aim, kinetics and thermodynamics of ^{125}I-β_2M binding to its receptors were studied by measuring the temperature dependence of the rate constants according to the classical approach for the study of molecular interactions.

Materials and Methods

4.1.Materials

4.1.1.Chemicals

All chemicals and reagents previously mentioned in section (2.1.1) in chapter two were used in the experiments of this chapter.

4.1.2.Instruments

The instruments previously mentioned in section (2.1.2) were used in the experiments of this chapter.

4.1.3.Samples

The tissue homogenates (ThI,ThII and ThIII) previously prepared in section (2.2.4) were included in this study as well as the two isolated forms of β_2M receptors (protein I and II)obtained in chapter three.

4.2.Methods

4.2.1.Time Course of ^{125}I-β_2M Binding to its Receptors in Breast Tumor Homogenates

Reagents

Tris buffer solutions (0.05 M) were prepared according to section (2.2.7). The pH was adjusted to the optimum value for each group (i.e. ThI, ThII and ThIII).

Procedure

1. The reaction mixture was prepared by mixing the reactants at their optimum concentrations (13.76×10^{-5}M of tracer and 300mg protein of ThII and 11.79×10^{-5} M of tracer with 400 mg protein of both ThI and ThIII).

2. The final volume was made up to 500 μl with tris buffer solution of the optimum pH value (7.4 for ThI and ThII and 7.6 for ThIII).

3. The assay tubes were incubated at 25°C for different time intervals (30, 45, 60, 75 and 90) minutes.

4. Steps 8-11 previously mentioned in section (2.2.7) in chapter two were repeated.

5. To determine the time course of ^{125}I-$β_2M$ binding to its receptors at different temperatures, steps 1, 2, 3 and 4 mentioned above were followed at different temperatures (5, 37 and 45) °C.

Calculations

1. The percent of specific binding was calculated as described in section (2.2.7).

2. The values of SB% were plotted against the corresponding time of incubation at each temperature.

4.2.2. Time Course of ^{125}I-$β_2M$ Binding to the Isolated Receptors

Reagents

Tris buffer solution (0.05M) of pH7.4 was prepared as described in section (2.2.7).

Procedure

1. A concentration of $9.83 \times 10^{-5}M$ of ^{125}I-$β_2M$ was added to 40 mg protein of both forms of receptors (protein I and II).

2. The final volume was made up to 500 μl with tris buffer solution pH 7.4. The assay tubes were incubated at 25°C for different time intervals (30, 45, 60, 75, and 90) minutes.

3. Steps 8-11 previously mentioned in section (2.2.7) were repeated.

4. To determine the time-course of the binding at different temperatures, steps 1, 2 and 3 mentioned above were followed at different temperatures (5, 37 and 45) °C

Calculations

1. The percent of specific binding was calculated as described in section (2.2.7) in chapter two.

2. SB% values were plotted against the corresponding time of incubation at each temperature.

4.2.3. Determination of Kinetic Parameters of ^{125}I-β_2M Binding to its Receptors in Breast Tumor Homogenates

Reagents

Tris buffer solution (0.05M) of optimum pH value was prepared as described in section (2.2.7).

Procedure

1. Increasing volumes (20, 30, 40, 50, 60, and 70) µl of ^{125}I-β_2M were added to 100 µl of each homogenate (ThI, ThII and ThIII) containing the optimum protein amount.

2. The final reaction volume was made up to 500µl with tris buffer solution of the optimum pH value for each group.

3. The assay tubes were incubated for 90 minutes at 25°C.

4. Steps 8-11 mentioned in section (2.2.7) were followed.

5. The above steps were repeated for each group at different temperatures (5, 37, and 45) °C.

Calculations

1. Scatchard plot was used to determine the kinetic parameters of the binding as follows:

$$\frac{B}{F} = \frac{1}{k_d}.B + \frac{B_{max}}{k_d}$$

Where:

B: ratio of specific binding.

F: concentration of free radioligand.

[F] could be determined as follows:

$$[F] = \frac{Total\ cpm - Bound\ cpm}{Total\ cpm} \times Concentration\ of\ tracer$$

B_{max}: maximal binding capacity.

K_d: dissociation constant.

K_a: affinity constant

2. The plot of B/F values against the corresponding values of B, gives linear relationship. B_{max} was calculated from the intercept on the x-axis, while the affinity constant was determined from the slope of the straight line.

4.2.4. Determination of Kinetic Parameter of ^{125}I-β_2M Binding to its Isolated Receptors

Reagents

Tris buffer solution (0.05M) pH 7.4 was prepared as previously described in section (2.2.7).

Procedure

1. The reaction mixture was prepared by incubating increasing volumes (10, 20, 30, 40, and 50) μl of ^{125}I-β_2M with 100 μl of the isolated receptors containing 40μg protein.

2. All tubes were incubated for 90 minutes at 25°C in a total volume of 500 μl and pH 7.4.Then steps 4 and 5 mentioned in section (4.2.3) were repeated.

Calculation

The calculations outlined in section (4.2.3) were followed to determine the values of K_a and B_{max} at each temperature.

4.2.5. Estimation of Hill Coefficients

Reagents

Tris buffer solutions of optimum pH values were prepared as previously mentioned in section (2.2.7).

Procedure

To estimate Hill coefficients, the reactants were incubated at their optimum conditions in a total volume of 500µl, then the steps 3, 4 and 5 described in section (4.2.3) were repeated.

Calculations

1. Counts of specific binding were calculated as mentioned in section (2.2.7).

2. The concentration of specifically bound $^{125}I\text{-}\beta_2M$ (B) was calculated as follows:

   $$B = \frac{\text{cpm of SB}}{\text{total cpm}} \times \text{total concentration of } ^{125}I\text{-}\beta_2M \text{ in incubation medium.}$$

3. Hill coefficients were obtained using the following equation; know as logarithmic form of Hill-equation [136].

$$\log \frac{B}{B_{max} - B} = n \log F - \log K'$$

Where:

F: free $^{125}I\text{-}\beta_2M$ in the incubation medium.

n: Hill coefficient.

K': a constant comprising the interaction factors and the intrinsic dissociation constant.

4. The values of log $B/(B_{max}-B)$ were plotted against log F values. Hill coefficient (n) was determined from the slope of the resulting straight line.

4.2.6. Thermodynamic Studies of ^{125}I-β_2M Binding to its Receptors

All data were obtained from the experiments mentioned in the previous sections of this chapter (4.2.1), (4.2.2), (4.2.3) and (4.2.4).

Calculations

1. The thermodynamic parameters of standard state were obtained from Van't Hoff plot, the values of the natural logarithm of affinity constant (Ka) obtained at different temperatures were plotted against the reciprocal values of the absolute temperature in Kelvin (1/T), according to the following equation:

$$\ln k_a = \frac{-\Delta H^o}{R} \cdot \frac{1}{T} + \text{Constant}$$

Where:

ΔH^o: the enthalpy change of the standard state.

ΔS^o : the entropy change of the standard state.

R : gas constant (8.314 J.K^{-1}.mol^{-1})

ΔH^o values were obtained from the slope of linear relationship of the plot. In this relation, an approximation was used to express that the standard reaction enthalpy are independent of temperature over the used range (5-45°C).

The Gibbs free energy (ΔG^o) for the binding reaction was determined from the following equation:

$$\Delta G^\circ = -RT \ln K_a$$

The entropy change of the standard state (ΔS°) at each temperature was calculated from the equation:

$$\Delta S^\circ = \frac{\Delta H^\circ - \Delta G^\circ}{T}$$

2. The thermodynamic parameters of the transition state were obtained from Arrhenius plot of $\ln k_{+1}$ values against the corresponding ($1/T$) values that gives linear relationship according to the following equation:

$$\ln k_{+1} = \ln A - \frac{E_a}{R} \cdot \frac{1}{T}$$

Where:

A: Arrhenius constant.

The activation energy (E_a) of the binding reaction was calculated from the slope of the straight line. The enthalpy of the transition state $\Delta H^\ast$ was calculated from the equation:

$$\Delta H^\ast = E_a - RT$$

The free energy change of the transition state $\Delta G^\ast$ was determined from the following equation:

$$\Delta G^\ast = -RT \ln k_{+1} + RT \ln \frac{KT}{h}$$

Where:

K: Boltzman constant (1.38×10^{-23} J.K^{-1}).

h : Plank constant(6.62×10^{-34} J.sec^{-1}).

The change in entropy of the transition state $\Delta S^\ast$ was calculating from the following formula:

$$\Delta S^\ast = \frac{\Delta H^\ast - \Delta G^\ast}{T}$$

Result and Discussion

4.3. Determination of the Kinetic Parameters

In receptor binding studies, the simplest assumption concerning the interaction of a radioactive ligand (^{125}I-β_2M) with the receptor (R) is the formation of the complex (^{125}I-β_2M/receptor) according to the following model:

$$^{125}I\text{-}\beta_2M + Receptor \rightleftharpoons (^{125}I\text{-}\beta_2M \text{ / Receptor})$$

Where:

k_{-1}: the rate constant of ^{125}I-β_2M association with the receptor.

k_{-1}: the rate constant of the complex dissociation.

At equilibrium:

$$K_a = \frac{[^{125}I - \beta_2M / Receptor]}{[^{125}I - \beta_2M][Receptor]} \tag{2}$$

$$K_a = \frac{1}{k_d} = \frac{k_{+1}}{k_{-1}} \tag{3}$$

Where:

K_a: the equilibrium constant of the association (affinity constant).

K_d: the equilibrium constant of the dissociation.

Scatchard equation was used to determine the affinity constants and maximum binding capacity at four different temperatures for all studied groups as well as the isolated β_2M receptors. Figure (4.1) and (4.2) show Scatchard plots for the studied groups and isolated receptors respectively.

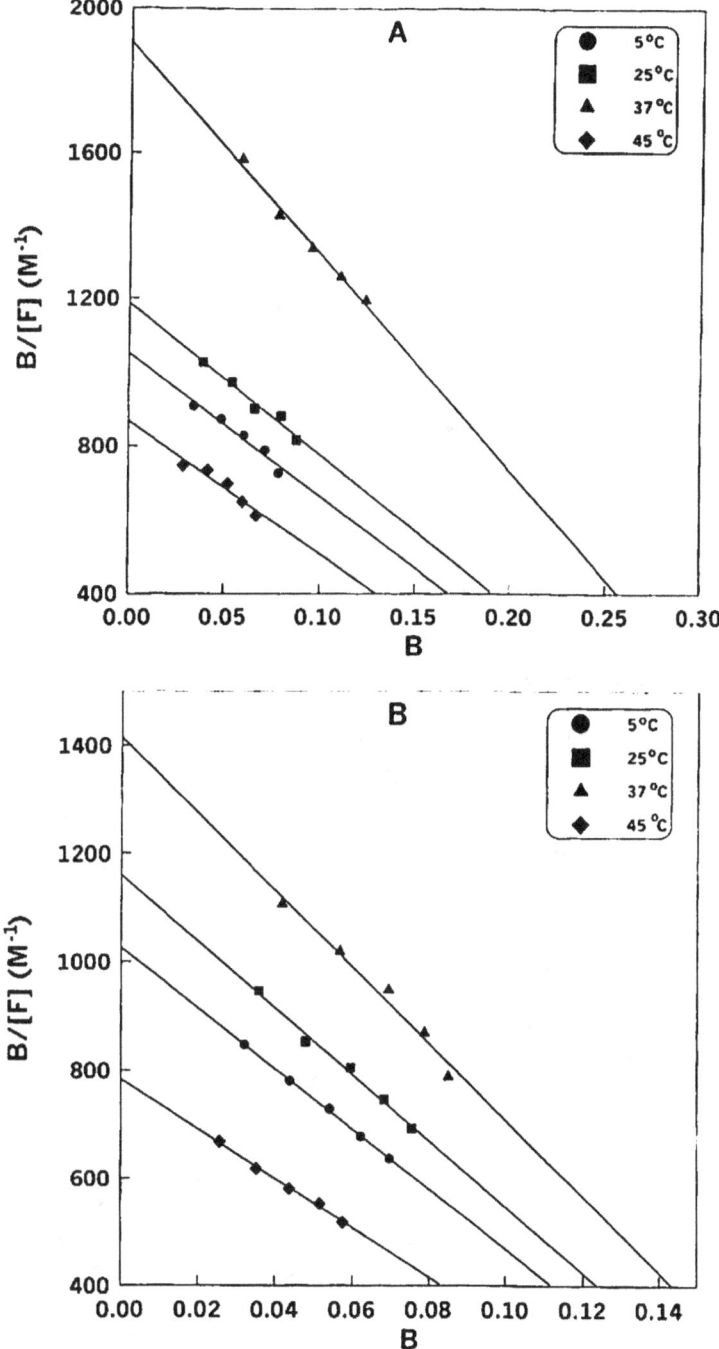

Figure (4.1). Scatchard plot for, A) ThI, B) ThII and C) ThIII at different temperatures. All details are described in section (4.2.3).

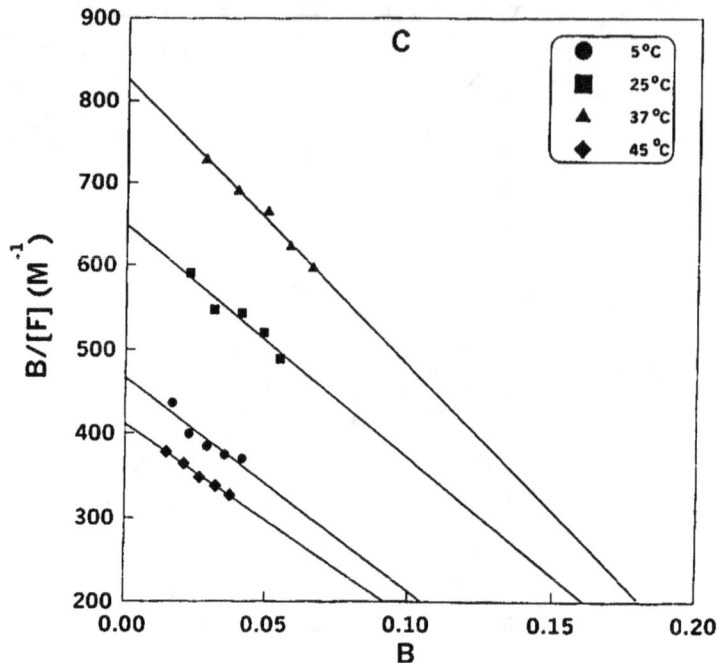

Figure (4.1). Continued.

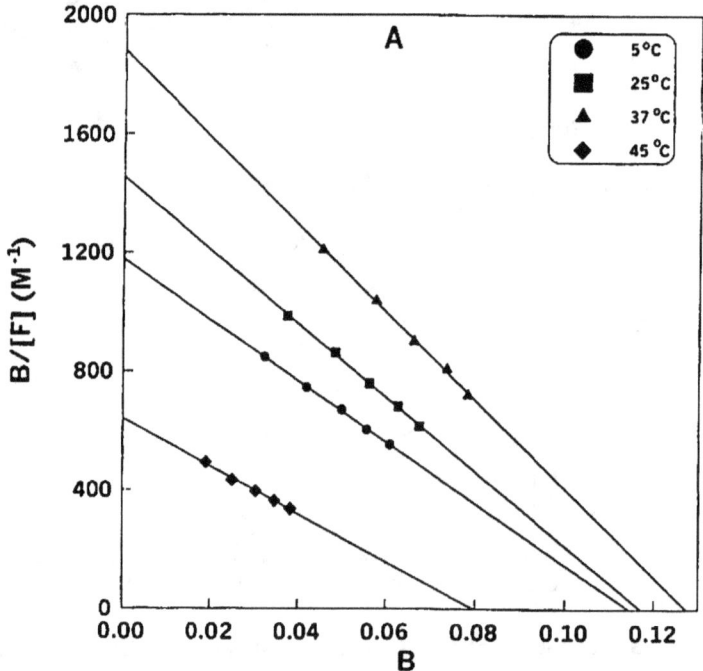

Figure (4.2). Scatchard plot for the isolated receptors at different temperatures, A) Protein I, B) Protein II. All details are described in section (4.2.4).

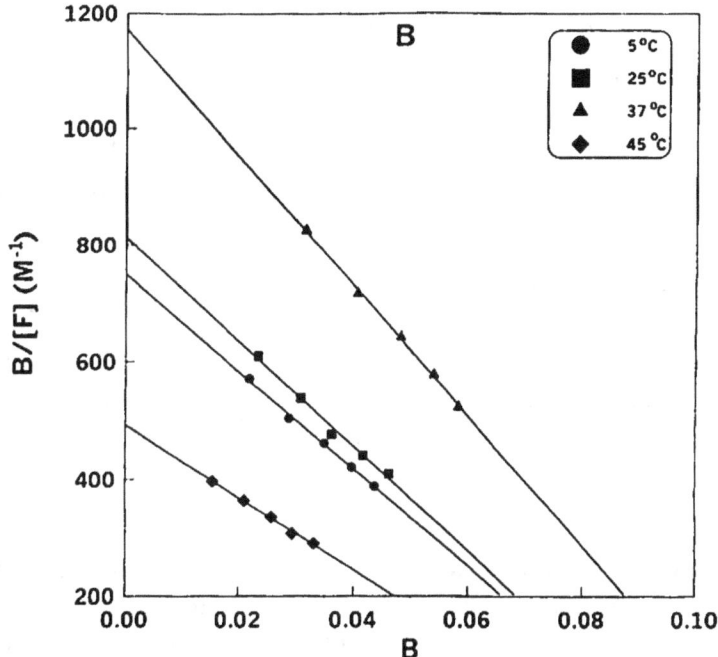

Figure (4.2). Continued.

The kinetic parameters (K_a) and (B_{max}) obtained in these studies are shown in table (4.1) and (4.2). The results reveal an increase in K_a and B_{max} values with increasing temperature in all groups until reaching 37°C where the maximum value of K_a indicates the highest affinity of ^{125}I-β_2M and the receptor to each other. These results are in agreement with those observed previously. It is known that temperature supply the reacting protein molecules with kinetic energy needed for the binding reaction up to a point, raised temperature affects the regular three-dimensional structure necessary for binding activity, hence K_a values are the lowest ones at 45°C.

The affinity constants and binding capacities for the isolated β_2M receptors also show an increase in their values with increasing temperature. Protein I (the high molecular weight from) exhibits higher affinity toward β_2M than the second from, this can be obviously seen in table (4.2), and could be an explanation for the greater values of SB% observed throughout this study. Also, K_a values are higher for protein I than ThI from which it was isolated and this

85

may be attributed to the isolation process, where the probability of ^{125}I-β_2M binding to its receptors was increased.

B_{max} values show insignificant differences within each type of the studied samples indicating that different temperatures have no effect on the theoretical total number of binding sites.

Scatchard plots generated from data of all groups studied showed no curvature in the plotted lines where the data obey the straight-line equation with good regression value. These results suggest the β_2M receptor molecules have a single binding site or more than one site with identical affinities.

Table (4.1). Association, dissociation constants and binding capacity for ^{125}I-β_2M binding to its receptors in breast tumor homogenates at different temperatures. All details are described in section (4.2.3).

Temperature °C	K_a (ml.mg^{-1})	K_d (mg.ml^{-1})	B_{max} (mg.ml^{-1})
Th I			
5	0.47	2.1	0.25
25	0.52	1.9	0.26
37	0.56	1.6	0.27
45	0.40	2.5	0.23
Th II			
5	0.32	3.0	0.37
25	0.34	2.8	0.40
37	0.49	2.0	0.45
45	0.30	3.2	0.34
TH III			
5	0.21	4.6	0.25
25	0.23	4.2	0.32
37	0.29	3.4	0.33
45	0.19	5.1	0.24

Table (4.2). Association, dissociation constants and binding capacity for ^{125}I-β_2M binding to the isolated receptors (protein I and II) at different temperatures. All details are described in section (4.2.4).

Temperature °C	K_a (ml.mg^{-1})	K_d (mg.ml^{-1})	B_{max} (mg.ml^{-1})
Protein I			
5	0.87	1.14	0.132
25	1.05	0.95	0.136
37	1.25	0.79	0.147
45	0.68	1.46	0.092
Protein II			
5	0.57	1.73	0.118
25	0.75	1.32	0.105
37	0.93	1.06	0.122
45	0.52	1.90	0.092

The time course experiments were carried out to obtain data concerning the reaction order of ^{125}I-β_2M binding to its receptors as shown in Figures (4.3) and (4.4).

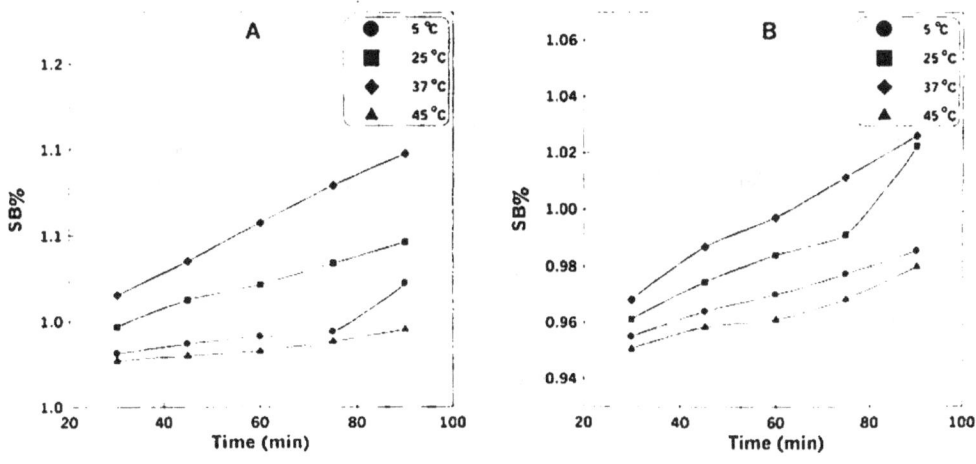

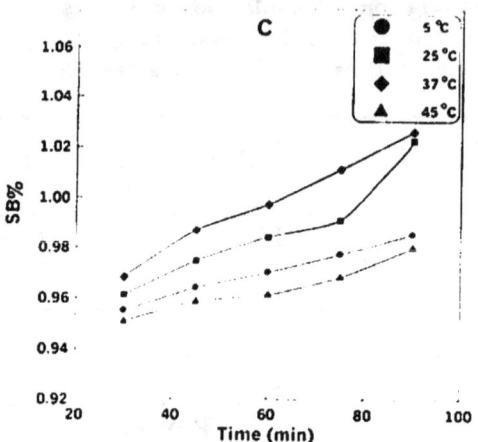

Figure (4.3). Time course of ^{125}I-β_2M binding to its receptors in breast tumor homogenates, A) ThI, B) ThII, C)ThIII. All details are described in section (4.2.1).

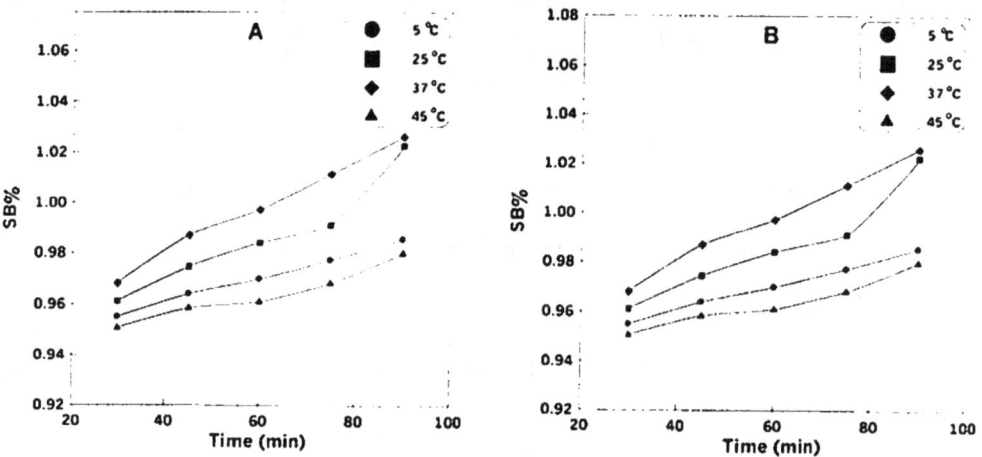

Figure (4.4). Time course of ^{125}I-β_2M binding to the isolated receptors, A) Protein I, B) Protein II. All details are described in section (4.2.2).

The graphical method was employed stating the rate law:

$$\text{Rate} = k_{+1} \, [^{125}\text{I-}\beta_2\text{M}][\text{Receptor}] \tag{4}$$

A first attempt was done assuming that $[^{125}\text{I-}\beta_2\text{M}] \gg [\text{Receptor}]$, and the resulting rate equation does fit the pseudo first order kinetics:

$$\text{Rate} = k' \, [^{125}\text{I-}\beta_2\text{M}] \tag{5}$$

And the rate law becomes:

$$\ln \frac{[^{125}\text{I} - \beta_2\text{M}]_\circ}{[^{125}\text{I} - \beta_2\text{M}]_\circ - [^{125}\text{I} - \beta_2\text{M} /\text{Receptor}]} = k'.t \tag{6}$$

Where:

k': the observed rate constant.

$[^{125}I\text{-}\beta_2M]_o$: the initial concentration of $^{125}I\text{-}\beta_2M$ at time 0.

$[^{125}I\text{-}\beta_2M\text{ /Receptor}]$: the concentration of the formed complex at time t.

Upon application of the available data, the resulting graph did not fit the law, where the regression of the straight line was very small for all the studied groups. A representative graph and the regression value are shown in figure (4.5). Consequently, the pseudo first order law was not used in determination of rate constant of $^{125}I\text{-}\beta_2M$ binding to its receptors.

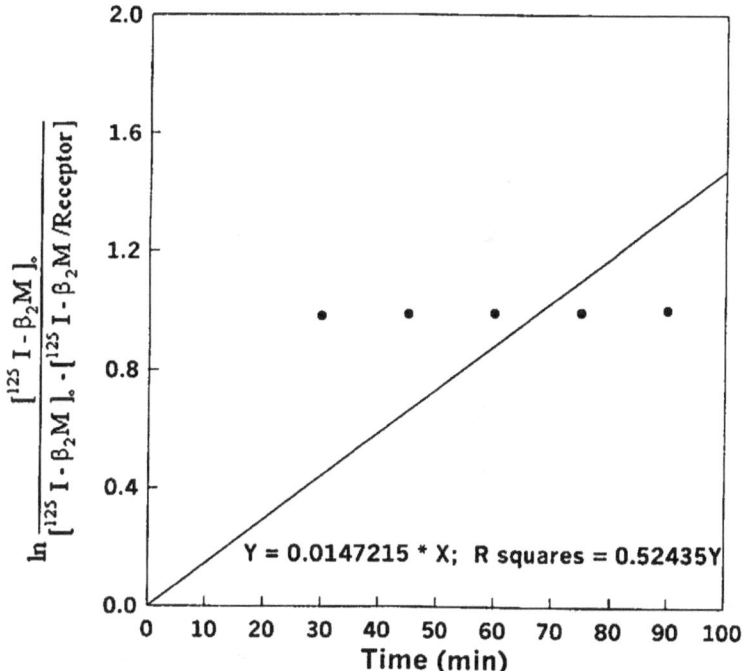

Figure (4.5). Representative figure for the estimation of first order rate constant for the studied groups. All other details are described in sections (4.2.1) and (4.2.2).

Second order kinetics were used according to the following mathematical formula:

$$\frac{1}{[^{125}I-\beta_2M]-[R]}.\ln\left(\frac{[^{125}I-\beta_2M]-[B_t]}{[R]-[B_t]}\right)=k_{+1}.t+\frac{1}{[^{125}I-\beta_2M]-[R]}.\frac{[^{125}I-\beta_2M]}{[R]} \quad (7)$$

Where:

[R]: Receptor concentration.

[B_t]: The concentration of (^{125}I-β_2M /Receptor) complex at time t.

k_{+1}: forward rate constant.

Figure (4.6) and (4.7) show the plot of the left-hand term of equation 7 against the time in minutes. The straight line has a good regression value for all types of samples studied. The slope of the resulting straight line equals to the forward rate constant ($k_{.1}$). The values of $k_{.1}$ at different temperatures were calculated using equation (3).

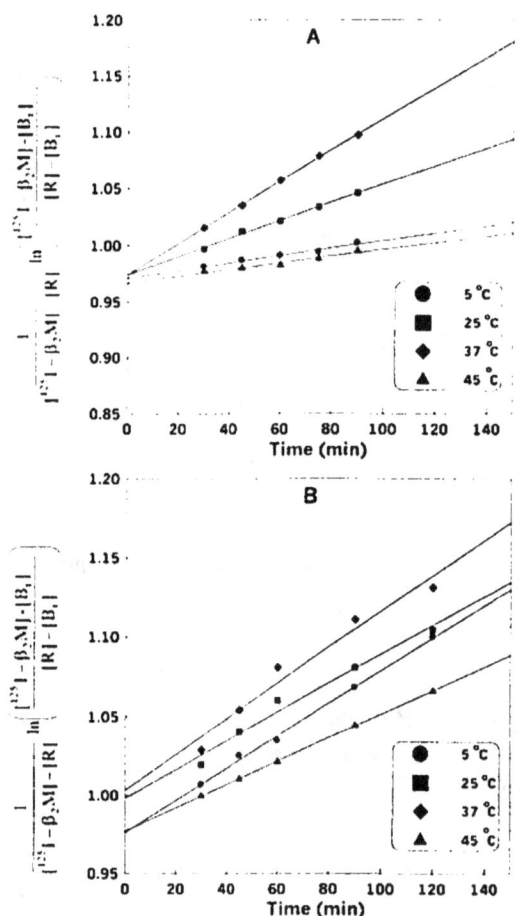

Figure (4.6). Kinetics of ^{125}I-β_2M binding to its receptors in breast tumor homogenates at different temperatures using second order rate law, A) ThI, B) ThII, C)ThIII. All details are described in section (4.2.1).

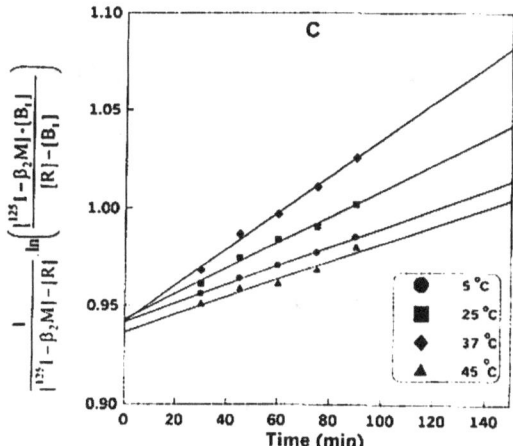

Figure (4.6). Continued.

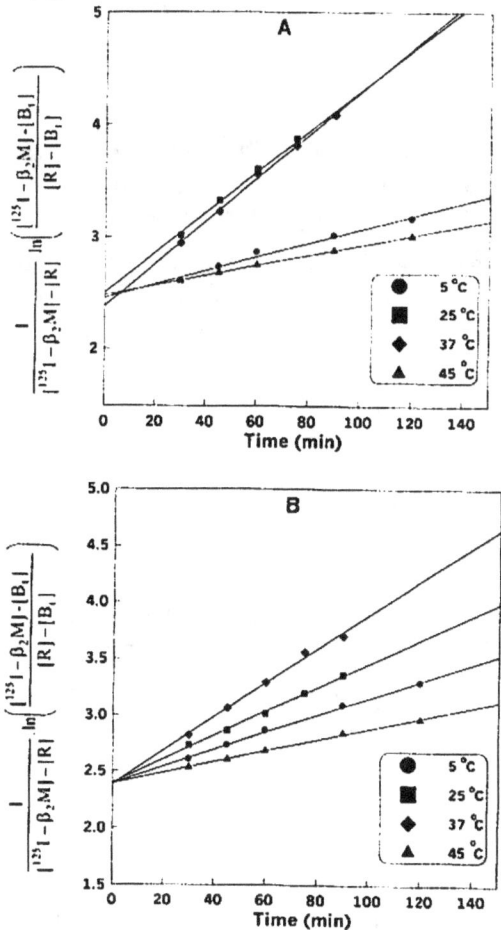

Figure (4.7). Kinetics of ^{125}I-β_2M binding to the isolated receptors at different temperatures using second order rate law, A) Protein I, B) Protein II. All details are described in section (4.2.2).

91

The half-life time of association, which represents the time needed for the formation of half amount of the complex at equilibrium, was determined, while the half-life time of dissociation ($t_{1/2 \text{ diss.}}$) was calculated from:

$$t_{1/2 \text{ diss.}} = \frac{1}{k_{+1} \times [^{125}I - \beta_2 M]}$$

Tables (4.3) and (4.4) show the kinetic parameters for the studied samples and the isolated receptors respectively. There was an increase in values of k_{+1} with increasing temperature. The results revealed that the rate of reaction is temperature dependent and k_{+1} value at 37°C is the highest among other values at 5, 25 and 45°C.

Table (4.3). Kinetics parameters for ^{125}I-β_2M binding to its receptors in breast tumor homogenates at different temperatures using second order rate law. All details are described in section (4.2.1).

Temperature °C	k_{+1} (mg^{-1}.ml.min^{-1})	k_{-1} (min^{-1}) x 10^{-3}	$t_{1/2 \text{ ass.}}$ (min.)	$t_{1/2 \text{ diss.}}$ (min.)
Th I				
5	0.0011	2.16	601	320
25	0.0013	2.68	441	258
37	0.0017	2.91	353	238
45	0.0007	1.90	831	365
Th II				
5	0.0003	1.15	984	692
25	0.0008	2.29	899	302
37	0.0013	2.77	521	250
45	0.0002	0.65	990	1060
Th III				
5	0.0004	2.25	1481	307
25	0.0006	2.66	1145	260
37	0.0009	3.19	768	217
45	0.0003	1.56	2366	443

Table (4.4). Kinetics parameters for^{125}I-β_2M binding to the isolated receptors at different temperatures using second order rate law. All details are described in section (4.2.2).

Temperature °C	k_{+1} (mg^{-1}.ml.min^{-1})	k_{-1} (min^{-1}) x 10^{-3}	$t_{1/2\ ass.}$ (min.)	$t_{1/2\ diss.}$ (min.)
Protein I				
5	0.0061	7.03	141	99
25	0.0180	17.1	48	41
37	0.0200	15.9	46	43
45	0.0045	6.61	192	105
Protein II				
5	0.0076	13.21	113	52
25	0.0106	14.06	81	49
37	0.0150	16.02	57	43
45	0.0048	9.27	177	75

β_2M receptors from ThI, ThII and ThIII exhibited different affinities towards β_2M, hence different rates of binding according to k_{-1} values; this may be attributed, presumably, to their site of origin.

The intercepts values obtained from the straight lines using second order rate law, figure (4.6), showed significant differences in the three studied samples (ThI,ThII and ThIII), while these values were almost the same in the two isolated forms (protein I and II) of β_2M receptors , figure (4.7). These observations indicate the high applicability of the rate law for the receptors after their isolation process resulting in low deviation errors of the data.

4.4.Determination of Hill Coefficient (n)

Figure (4.8) and (4.9) show the Hill plot of ^{125}I-β_2M binding to its receptors. Determination of Hill coefficient can provide information about the extent to which the ligand binding sites are cooperative and the effect of temperature on their cooperativity. The value of Hill coefficient cannot exceed the number of ligand binding sites per receptor molecule making its determination of value in estimating their number.

The values of Hill coefficients obtained in this study are shown in table (4.5) and (4.6) which reveal that Hill coefficients are almost constant indicating that the cooperativity of ^{125}I-β_2M binding sites are slightly affected by temperature.

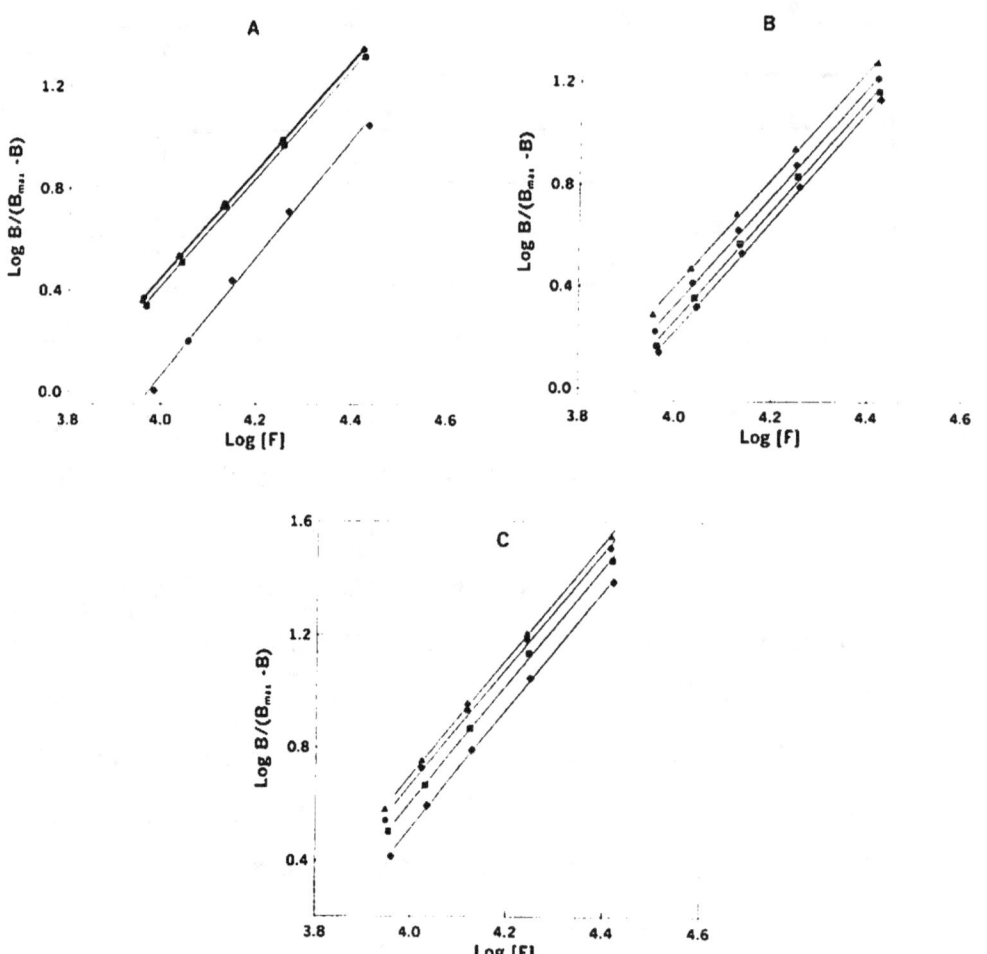

Figure (4.8). Determination of Hill coefficient of ^{125}I-β_2M binding to its receptors in breast tumor homogenates at different temperatures, A) ThI, B) ThII, C) ThIII. All details are described in section (4.2.5).

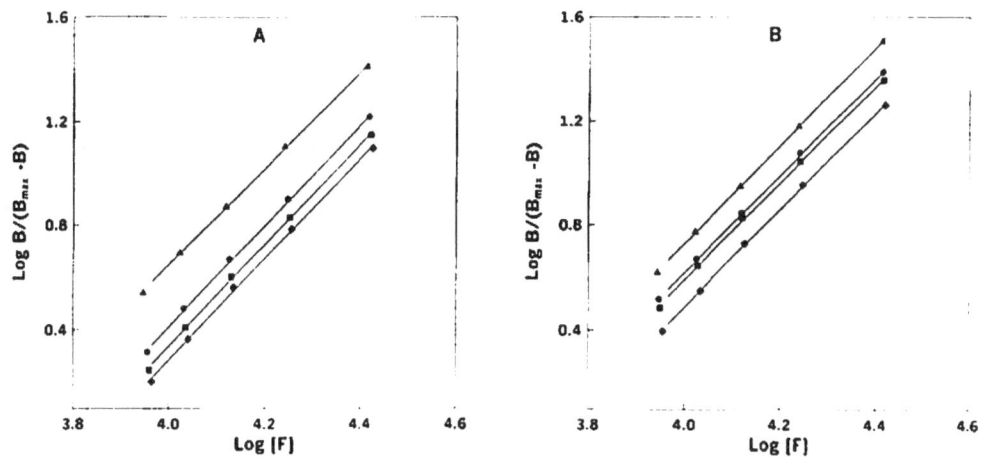

Figure (4.9). Determination of Hill coefficient of ^{125}I-β_2M binding to the isolated receptors at different temperatures, A) protein I, B) protein II. All details are described in section (4.2.5).

Table (4.5). Hill coefficients of ^{125}I-β_2M binding to its receptors in breast tumor homogenates at different temperatures. All details are described in section (4.2.5).

Temperature (°C)	Hill Coefficient (n)		
	ThI	ThII	ThIII
5	2.14	2.13	2.06
25	2.16	2.14	2.09
37	2.16	2.21	2.10
45	2.12	2.13	2.07

Table (4.6). Hill coefficients of ^{125}I-β_2M binding to the isolated receptors at different temperatures. All details are described in section (4.2.5).

Temperature (°C)	Hill Coefficient (n)	
	Protein I	Protein II
5	1.93	1.97
25	1.93	1.96
37	1.92	1.96
45	1.89	1.98

4.5.Determination of Thermodynamic Parameters

A. Thermodynamic Parameters of Standard State

Figure (4.10) and (4.11) represent the relation between the equilibrium binding constant (affinity constant) for ^{125}I-β_2M binding to its receptors on temperature.

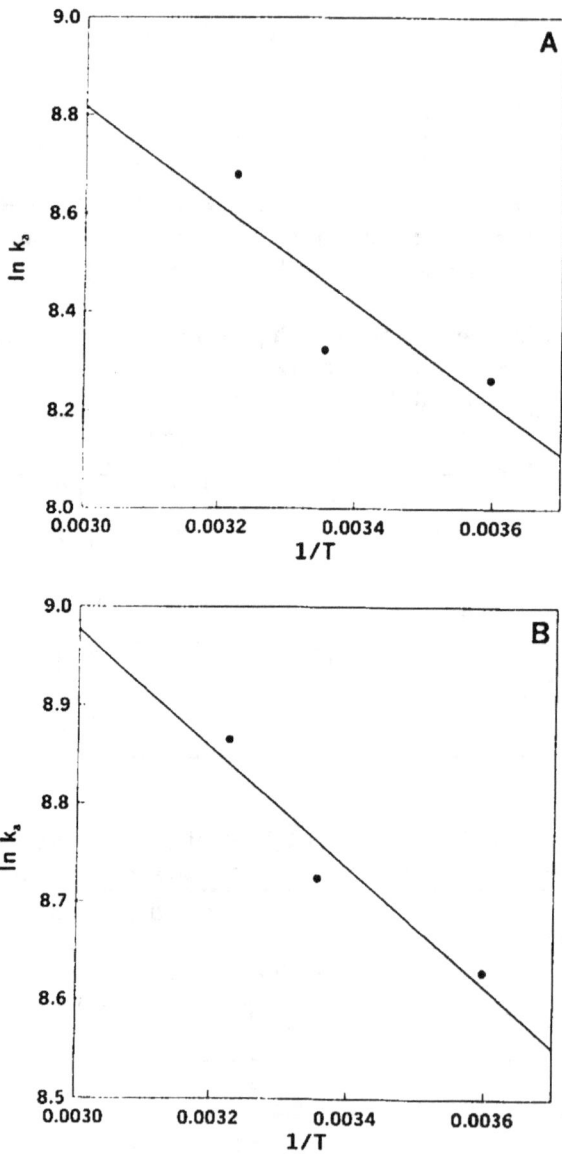

Figure (4.10). Relation between lnK_a and 1/T for ^{125}I-β_2M binding to its receptors in breast tumor homogenates at different temperatures. A) ThI, B) ThII, C) ThIII. All details are described in section (4.2.6).

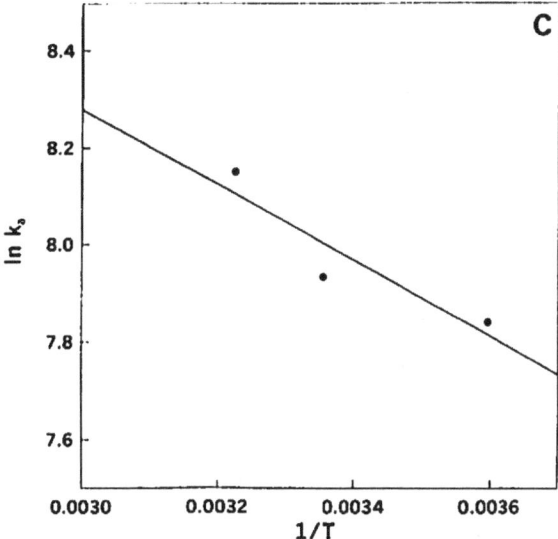

Figure (4.10). Continued

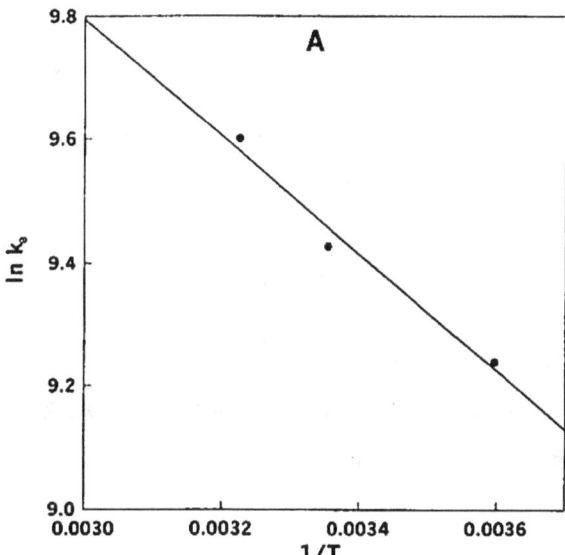

Figure (4.11). Relation between $\ln K_a$ and $1/T$ for ^{125}I-β_2M binding to the isolated receptors different temperatures, A) Protein I, B) Protein II. All details are described in section(4.2.6).

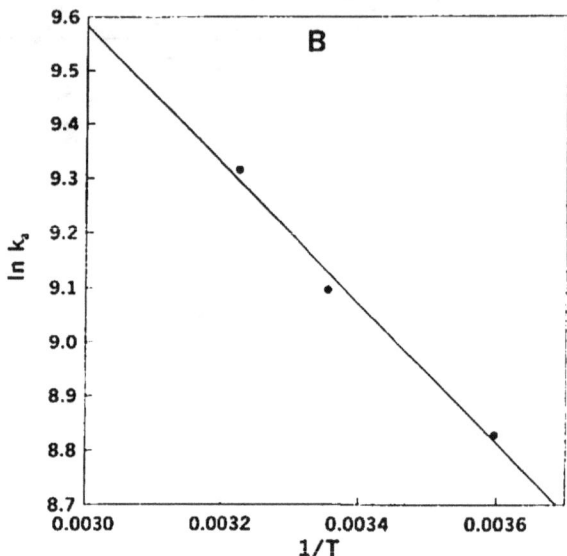

Figure (4.11). Continued.

The results obtained in this study, assuming that ΔH value is constant over the used range of temperature, revealed that ΔH^o had positive values for all studied samples (ThI, ThII and ThIII) and the isolated receptors indicating that the binding reaction is an endothermic process and implying the favorable non-covalent interaction between ^{125}I-β_2M and the receptors.

The negative values of ΔG^o observed in tables (4.7) and (4.8) indicate the spontaneity of the binding reaction. There were no significant differences in ΔG^o values obtained from the binding of ^{125}I-β_2M to its receptors from three types of tissue homogenate (ThI, ThII and ThIII) and the binding reaction seems, generally, to proceed nearly less spontaneously than the case of the reaction of the isolated receptors.

The results also show that the ΔS^o values decrease with increasing temperature. This may reflect the stability and the highly arranged status of (^{125}I-β_2M/receptor) complex at 37°C. A positive value of ΔS^o obtained upon reaction of all samples, could suggest that the binding spontaneity was entropically driven including the hydrophobic interactions that play an important role in stabilizing the complex [137].

Table (4.7). Thermodynamic parameters at the standard state for ^{125}I-β_2M binding to its receptors in breast tumor homogenates at different temperatures. All details are described in section (4.2.6).

Temperature $^\circ C$	ΔH° (KJ.mol^{-1})	ΔS° (J.mol^{-1}.K^{-1})	ΔG° (KJ.mol^{-1})
Th I			
5	50.46	90	-19.9
25	50.46	89	-21.6
37	50.46	90	-22.8
45	50.46	86	-22.3
Th II			
5	83.75	99	-19.1
25	83.75	97	-20.6
37	83.75	99	-22.4
45	83.75	94	-21.6
Th III			
5	64.73	88	-18.1
25	64.73	88	-19.7
37	64.73	89	-21.0
45	64.73	85	-20.5

Table (4.8). Thermodynamic parameters at the standard state for ^{125}I-β_2M binding to the isolated receptors at different temperatures. All details are described in section (4.2.6).

Temperature °C	ΔH^o (KJ.mol^{-1})	ΔS^o (J.mol^{-1}.K^{-1})	ΔG^o (KJ.mol^{-1})
Protein I			
5	78.85	117	-21.4
25	78.85	116	-23.4
37	78.85	116	-24.7
45	78.85	110	-23.8
Protein II			
5	106.9	113	-20.4
25	106.9	113	-22.5
37	106.9	113	-24.0
45	106.9	108	-23.1

B. Thermodynamic Parameters of Transition State

The transition state theory proposes a model in which the association of two proteins (ligand and the receptor) to form the final product (complex) proceeds through the formation of an activated complex (transition state)[138]. Consequently, the interaction of ^{125}I-β_2M with its crude and isolated receptors can be represented as follows:

$$^{125}\text{I-}\beta_2\text{M} + \text{Receptor} \rightleftharpoons (^{125}\text{I-}\beta_2\text{M / Receptor})^* \text{Complex}$$

Reactants   Activated Complex
(State I)   (State II)

$$(^{125}\text{I-}\beta_2\text{M / Receptor}) \text{ Complex}$$

Product
(State III)

100

State I represents the initial energy level of ^{125}I-β_2M and the receptor. In state II, the two reactants interact to form the activated complex (^{125}I-β_2M/recptor)*. State III describes the final fully interacting product (^{125}I-β_2M/receptor) complex.

The thermodynamic parameters of the transition state (ΔH^*, ΔG^* and ΔS^*) were calculated employing Arrhenius equation. Figures (4.12) and (4.13) show the plot of $\ln k_{+1}$ against ($1/T$) values in which the slope of the resulting straight line represents the activation energy (E_a).

Tables (4.9) and (4.10) show the values of the thermodynamic parameters of the transition state for ^{125}I-β_2M binding to its crude and isolated receptors respectively. The value of activation energy represents the required energy to overcome the energy barrier of the transition state for the formation of (^{125}I-β_2M /Recrptor) complex. The binding of ^{125}I-β_2M to its receptors in ThI seems to need more energy to reach the activated complex (state II), and also protein I (isolated form) as compared to protein II. The value of activation energy is in accordance with the positive value of ΔG^* which indicates that the process to reach the activated complex is a non-spontaneous one.

The results show that there was a slight decrease in ΔH^* values with increasing temperature which could be attributed to its relation to E_a according to the equation: $\Delta H^* = E_a - RT$, in which the numerical value of RT is small in comparison with Ea of the binding reaction.

The negative ΔS^* values obtained for all studied samples indicate that the activated complex involved in the binding reaction is of a more arranged structure than the starting reactant in the binding medium.

The values of the thermodynamic parameters obtained from the study of ^{125}I-β_2M binding to its receptors, give an idea concerning the nature of forces that hold the two proteins together to form the complex.

Blumenthal D. and Stull J.[139] have proposed that the formation of ligand-receptor complex occurs, firstly, by its stabilization by hydrophobic interactions and secondly is the stabilization with short range of interactions such as electrostatic interactions, hydrogen bonding and Van der Waals forces. Hydrophobic interactions contribute to the complex stability via high positive entropy change which means a value of ΔS^{*} greater than zero, while electrostatic interactions, hydrogen bonds and Van der Waals forces contribute to the complex formation by negative entropy change where ΔS^{*} value is less than zero [140].

Considering the negative values of ΔS^{*} obtained for the studied samples, the formed complex might be stabilized by electrostatic interactions, hydrogen bonds and Van der Waals forces rather than hydrophobic interactions, while the binding reaction is entropically driven.

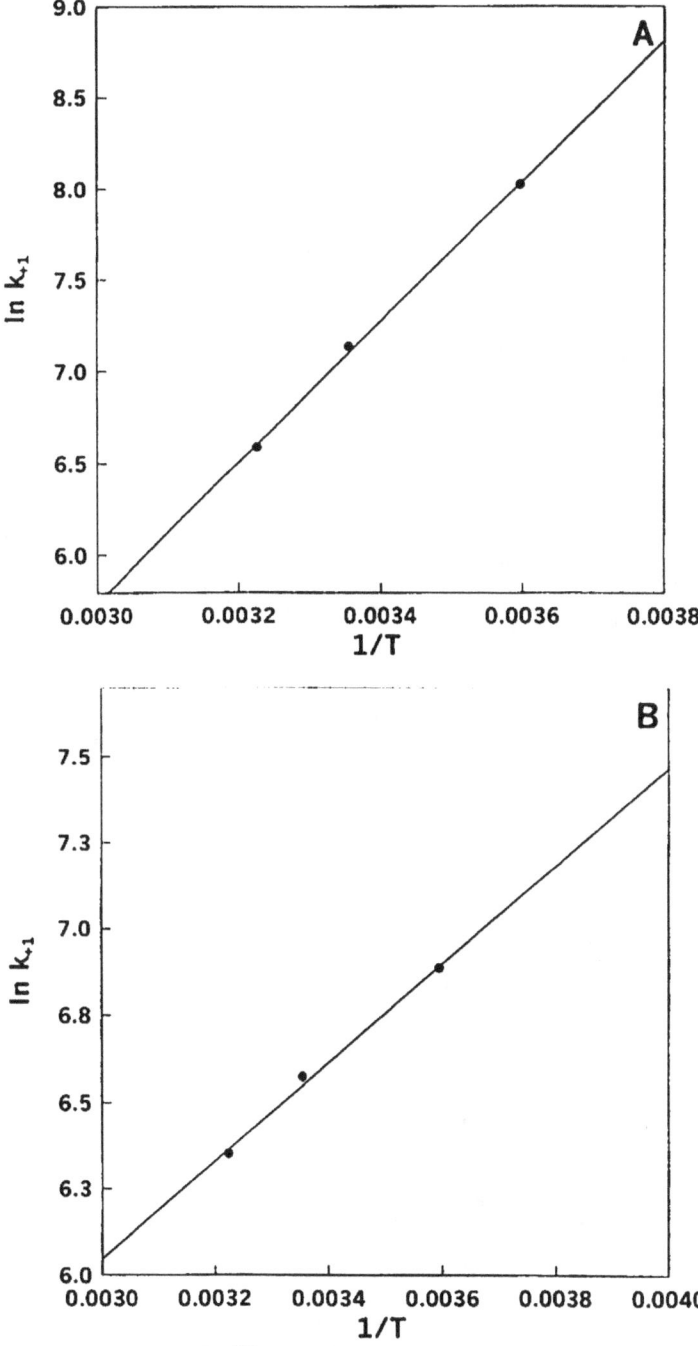

Figure (4.12). Arrhenius plot for ^{125}I-β_2M binding to its receptors in breast tumor homogenates at different temperatures. A) ThI, B) ThII, C) ThIII. All details are described in section (4.2.6).

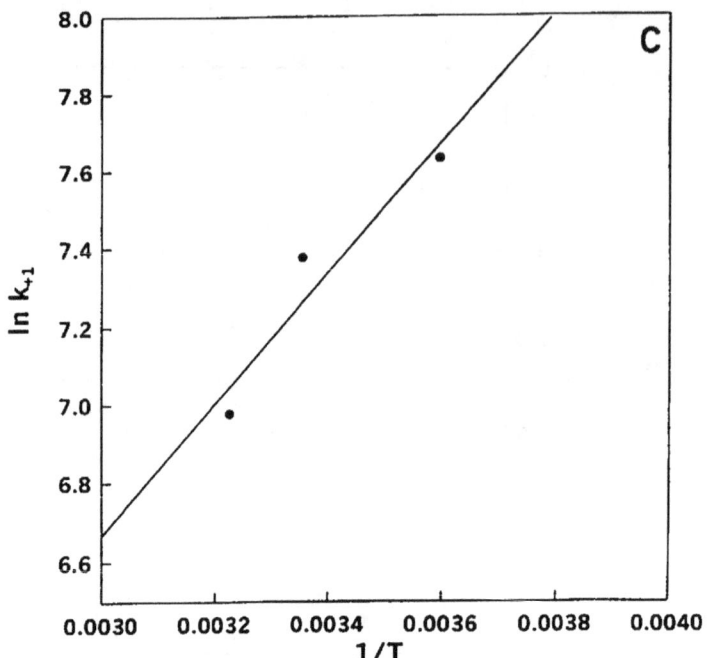

Figure (4.12). Continued.

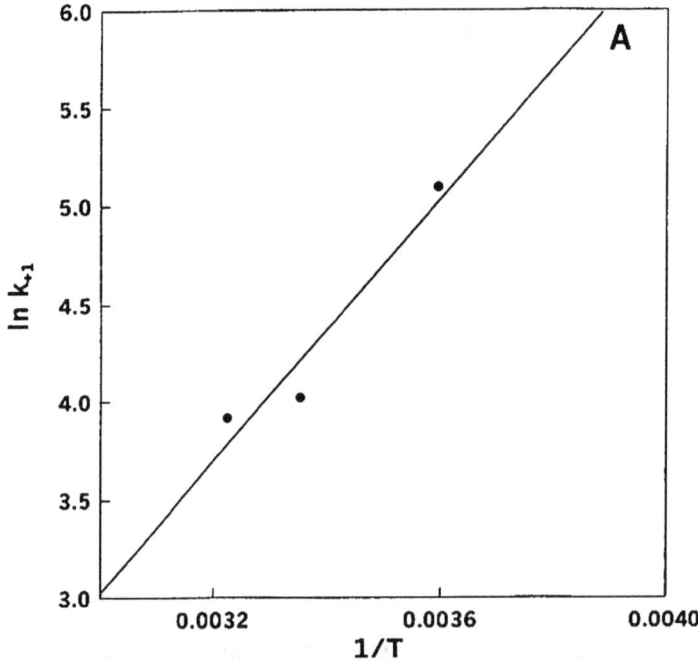

Figure (4.13). Arrhenius plot for ^{125}I-β_2M binding to the isolated receptors at different temperatures, A) Protein I, B) Protein II. All details are described in section (4.2.6).

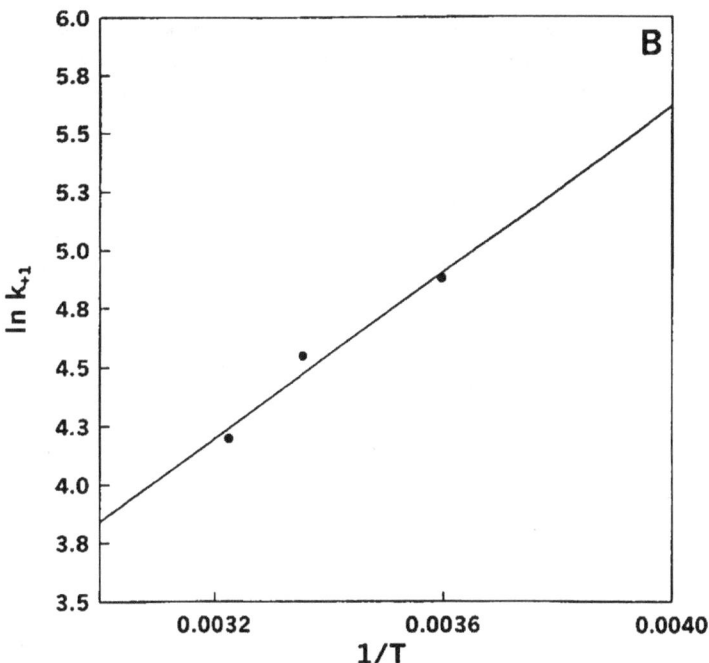

Figure (4.13). Continued.

Table (4.9). Thermodynamic parameters for the transition state for ^{125}I-β_2M binding to its receptors in breast tumor homogenates at different temperatures. All details are described in section (4.2.6).

Temperature °C	E_a (KJ.mol^{-1})	ΔH^* (KJ.mol^{-1})	ΔS^*(J.mol^{-1}.K^{-1})	ΔG^* (KJ.mol^{-1})
Th I				
5	11.7	9.5	-268	84
25	11.7	9.3	-268	89
37	11.7	9.2	-268	92
45	11.7	9.1	-277	97
Th II				
5	12.6	10.4	-274	86
25	12.6	10.2	-270	91
37	12.6	10.1	-267	93
45	12.6	10.0	-285	101
Th III				
5	13.9	11.6	-266	86
25	13.9	11.5	-268	91
37	13.9	11.4	-267	94
45	13.9	11.3	-277	99

Table (4.10). Thermodynamic parameters for the transition state for ^{125}I-β_2M binding to the isolated receptors at different temperatures. All details are described in section (4.2.6).

Temperature °C	E_a (KJ.mol^{-1})	ΔH^* (KJ.mol^{-1})	ΔS^* (J.mol^{-1}.K^{-1})	ΔG^* (KJ.mol^{-1})
Protein I				
5	27.8	25.5	-195	80
25	27.8	25.3	-193	83
37	27.8	25.2	-196	86
45	27.8	25.2	-211	92
Protein II				
5	14.7	12.4	-240	79
25	14.7	12.3	-242	84
37	14.7	12.2	-241	87
45	14.7	12.1	-252	92

Spectroscopic Studies on β$_2$M and (^{125}I-β$_2$M /Receptor) Complex

Chapter Five

Spectroscopic Studies on β₂M and (¹²⁵I-β₂M/Receptor) Complex

Introduction

Spectroscopy is one of the most valuable techniques available in biochemistry. Protein molecules may be identified by their characteristic absorption spectra in the ultraviolet, visible or infrared[141].

The wavelength range (210-300 nm.) is the most important region in characterizing protein molecules. Absorption in this region is related to tryptophan, tyrosine, phenylalanine and histidine amino acid residues other amino acids show only insignificant absorption spectra [142].

A molecule or part of a molecule that can be existed by absorption is called a chromophone. The absorption spectrum of a chromophore is primarily determined by the chemical structure of the molecule. However, a large number of environmental factors produce detectable changes in spectroscopic parameters such as λ_{max} [143] . Environmental factors consist of pH, the polarity of the solvent or neighboring molecules and the relative orientation of neighboring chromophores. It is precisely these environmental effects that provide the basis for the use of absorption spectroscopy in characterizing macromolecules [143].

There have been no similar studies dealing with the spectroscopic behavior of β₂M, the receptors or their binding reaction, hence this chapter tries to shed light on the spectroscopic aspects of standard β₂M equipped with its kit and the complex obtained from the reaction with its receptors.

Materials and Methods

5.1. Materials

5.1.1.Chemicals

All chemicals and regents mentioned in section (2.1.1) were used in the experiments of this chapter.

5.1.2.Instruments

All instruments mentioned in section (2.1.2) were used in the experiments of this chapter, in addition Double Beam Spectrophotometer, Cintra 5 (Australia).

5.1.3.Buffers and Samples

Tris buffer solutions (0.05 M) pH 7.4 were prepared as previously described in section (2.2.7) and were used in the experiments of this chapter.

The sample denoted (protein I) was used in the experiments of this chapter as a source of $\beta_2 M$ receptors.

5.2. Methods

5.2.1.Preparation and Gel Filtration of (^{125}I-$\beta_2 M$/Receptor) Complex

Procedure

1. ^{125}I-$\beta_2 M$ was reacted with its receptors (protein I) at the previously determined optimum conditions.

2. At the end of the reaction, 555μl of the reaction mixture were applied to the surface of gel chromatography column, equilibrated with Tris buffer pH 7.4 at a flow rate of 10ml/ hr.

3. Fractions of 1 ml were collected; the radioactivity of each fraction was counted.

5.2.2. The UV Spectra of β_2M and (^{125}I-β_2M/Receptor) Complex

Procedure

1. One hundred microlitters of standard β_2M were mixed with 400 µl of tris-buffer solution of pH 7.4 and placed in a 0.5 cm^3 quartz cuvette.

2. The absorption spectrum of this sample was scanned in the UV region against an appropriate blank.

3. The same steps were repeated for the free receptors (protein I) and (^{125}I-β_2M/ receptor) complex prepared in the previous section.

5.2.3. Factors Affecting the Absorption Properties of β_2M and (^{125}I-β_2M/Receptor) Complex

The Effect of pH

Regents

Tris buffer solution (0.05 M) of different pH values 4, 7.4 and 9 and 11 was prepared as described in section (2.2.7). The pH was adjusted by using 1N HCl solution.

Procedure

1. One hundred microlitters of standard β_2M were mixed with 400µl of tris buffer of pH (4, 7.4, 9 and 11) in a 0.5cm^3 quartz cuvette.

2. The sample was scanned in UV region against a buffer blank at each pH.

3. Both steps were repeated for the free receptor (protein I) and (^{125}I-β_2M/receptor) complex.

Effect of Solvent Polarity

1. One hundred microlitters of $\beta_2 M$ were separately mixed with 200μl of (ethanol, ethylene glycol, glycerol and dioxane).

2. The final volume was made up to 1000 μl with tris-buffer pH 7.4.

3. The absorbance of each sample was scanned in the UV region against a blank reference contains 20% solvent in tris-buffer 7.4.

4. The steps 1,2 and 3 were repeated for the free receptors (protein I) and (^{125}I-$\beta_2 M$/receptor) complex.

5.2.4. Spectrophotometric pH Titration of $\beta_2 M$ and (^{125}I-$\beta_2 M$/Receptor) Complex

Reagents

Tris buffer solutions of different pH values were prepared as described in section (2.2.7) and the desired pH was adjusted using HCl solution.

Procedure

1. One hundred microlitters of $\beta_2 M$ were mixed with 400μl of a series of buffers at pH ranging from 4 to 8.

2. The absorbance of each pH was recoded at 211 nm. The step was repeated for another series at pH range (8-12) and 295 nm.

3. The absorbance values were plotted at the corresponded pH values.

4. Steps 1,2 and 3 were repeated for (^{125}I-$\beta_2 M$/receptor) complex.

Results and Discussion

5.3. The UV Spectra of β_2M and (^{125}I-β_2M/Receptor) complex

The absorption spectra of standard β_2M was scanned in the UV region. Figure (5.1) shows the UV spectrum of β_2M at pH 7.4. Two strong peaks were observed, the first at 240 nm and the second at 273 nm. Both peaks are related to tyrosine residues [144], which may be located on the surface of the β_2M molecule.

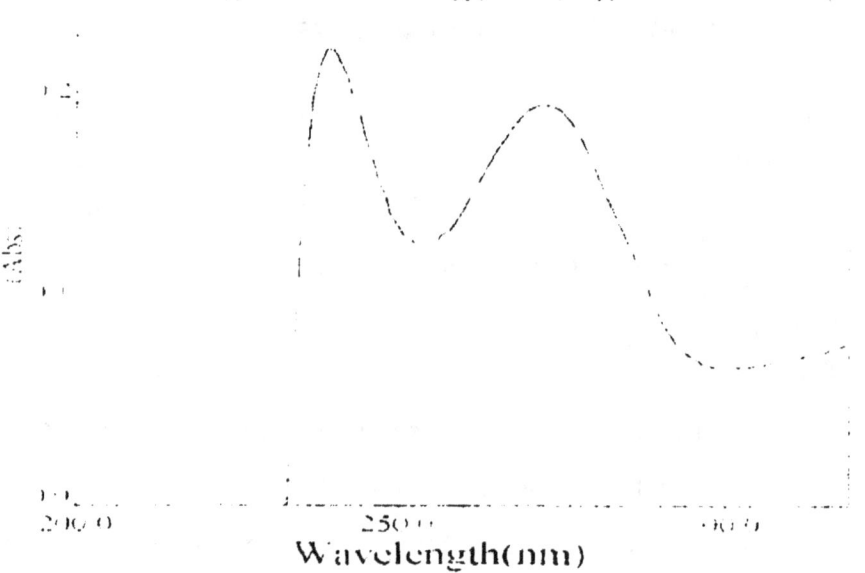

Wavelength(nm)

Figure (5.1). UV spectrum of ^{125}I-β_2M at pH 7.4. All details are described in section (5.2.2).

Figure (5.2) shows the UV spectra of the complex (^{125}I-β_2M/receptor) prepared by incubating ^{125}I-β_2M with protein I sample. The figure reveals a strong peak at 231 nm, which may be related to tyrosine residues, the observed blue shift (to a less λ_{max} value) could be attributed to the conformational changes that may occur within the two protein molecules upon binding resulting in alteration in the relative orientation of the neighboring groups. Figure (5.2) also shows a second peak, the appearance of this broad peak might be related to the absorption spectrum of tyrosine residues on β_2M molecule combined with that of phenylalanine residues, which may be positioned on the receptor molecule in contact with the medium.

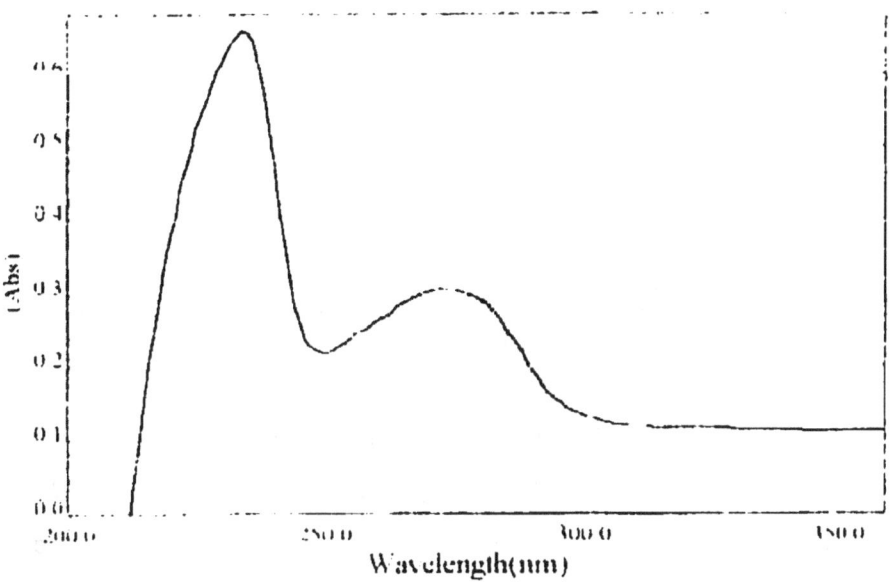

Figure (5.2). UV spectrum of (^{125}I-β_2M/receptor) complex at pH 7.4. All details are described in section (5.2.2).

The absorption spectra of the isolated receptors (protein I) are shown in figure (5.3) in which a single peak can be observed at 263 nm related to the absorption spectrum of phenylalanine [144], this result support the previous explanation for the broad peak in figure (5.2).

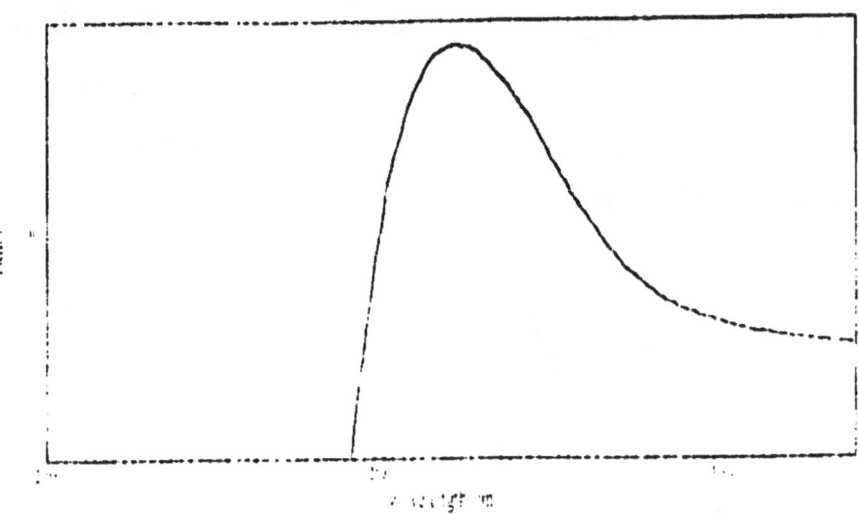

Figure (5.3). UV spectrum of β₂M receptors (protein I) at pH 7.4. All details are described in section (5.2.2).

The results obtained in this study indicate that the peaks observed for free β₂M in figure (5.1) and free receptor in figure (5.3) did not disappear due to binding, figure (5.2). It could be suggested [144] that amino acid residues of tyrosine and phenylalanine are not located in the interacting interface of the two proteins; hence their absorption spectra appeared only with some shifts in λ_{max}.

5.4. Factors Affecting the Absorption Properties
Effect of pH

The pH of medium determines the ionization state of ionizable chromophores. Table (5.1) illustrates the effect of pH on the λ_{max} of β₂M and (^{125}I-β₂M/receptor) complex. There was an increase in λ_{max} value with increasing in medium pH value. A red shift was observed in absorption of tyrosine residues. This effect can be attributed to the ionization of the phenolic OH group in the side chain of tyrosine residues, which enables the lone pair of electrons on oxygen atom to be excited easier at a lower energy resulting in red shift.

Only insignificant absorption was observed at pH 12 which may be a result for the denaturation of β_2M molecule.

The same effect (red shift) was observed on λ_{max1} value of (^{125}I-β_2M/receptor) complex, the red shift is related to the changes in the ionization state of tyrosine residues as explained above, while the second peak (λ_{max2}) remained broad with insignificant shift. The change in pH had no effect on the λ_{max} value of the receptor as shown in table (5.1). This result confirms that the absorption is related to phenylalanine amino acid residues where pH do not affect the ionization state of these residues and supports the explanation made for absorption of the complex.

Table (5.1). Effect of increasing pH values on the λ_{max} of β_2M, the complex (^{125}I-β_2M /receptor) and the receptor. All details are described in section (5.2.3).

pH	λ_{max} (nm)		
	β_2M	Complex	Receptor
4	238, 270	229, 266	262
7.4	240, 273	231, 267	263
9	246, 276	238, 267	262
12	-	-	-

Effect of Solvent Polarity

The effect of solvent polarity on the absorption spectra of β_2M, the receptors and the complex formed from their binding.

The solvent perturbation method is carried out by measuring the spectra of proteins in polar and non-polar solvents. This method may be useful in the determination of the location of some amino acid residues whether in the internal of the molecule or on its surface. The effect of some non-polar solvents could be investigated by using a mixture of water and non-polar solvents that reduced the polarity. Several spectral changes may obtained in the presence of these perturbants, such as the change in the position of λ_{max} or the appearance of new chromospheres on the surface of the complex [144].

Table (5.2) reveals that the value of λ_{max} for $n\rightarrow\pi^*$ transition in tyrosine residues occurs at a longer wavelength as the polarity of the solvent decreases in the sequence:

dioxane > ethanol > ethylene glycol > glycerol

The increase in λ_{max} (red shift) observed in this study is due to the decrease in the solvent polarity and can be attributed to the effect of hydrogen bonding between the solvent protons and the non-bonding lone pair on the oxygen atom which will be decreased, hence less energy will be sufficient for electronic transitions. The red shift in the λ_{max} of tyrosine residues makes the possibility for the presence of those residues on the external of β_2M.

The effect of solvent polarity on λ_{max} of the receptor was observed to be towards shorter wavelength for $\pi\rightarrow\pi^*$ transition occurring in phenylalanine residues. This type of electronic transition is the more common one in biological molecules. Also, no significant shift was observed in λ_{max2} of (^{125}I-β_2M/receptor) complex, then it could be said the phenylalanine residue responsible for this spectrum might be buried in the internal of complex upon binding where it is insensitive to change solvent polarity [144].

Table (5.2). Effect of solvent polarity on the λ$_{max}$ of β$_2$M, the complex (^{125}I-β$_2$M /receptor) and the receptor. All details are described in section (5.2.3).

Solvent	λ$_{max}$ (nm)		
	β$_2$M	Complex	Receptor
Ethanol	231, 280	226, 267	258
Ethylene Glycol	235, 278	228, 266	259
Glycerol	238, 275	229, 267	261
Dioxane	229, 281	225, 267	255

5.5. Spectrophotometric pH Titration of β$_2$M and (^{125}I-β$_2$M/Receptor) Complex

Many studies of protein structure require the determination of pKa alues for proton dissociation from ionizable amino acid side chains, because these values give an indication of the location of the amino acid in the protein. This can often be done spectrophotometrically because association often changes the spectrum of one of the chromophores.

Spectrophotometric titration was carried out for two amino acid residues, histidine at 211 nm and tyrosine at 295 nm for both standard β$_2$M and the (^{125}I-β$_2$M/receptor) complex. The values of titration gave an indication for their positions in the protein and gave a simple picture about protein conformation.

Figure (5.4) shows the titration curves for the β$_2$M and the (^{125}I-β$_2$M/receptor) complex at 211 nm for histidine residue.

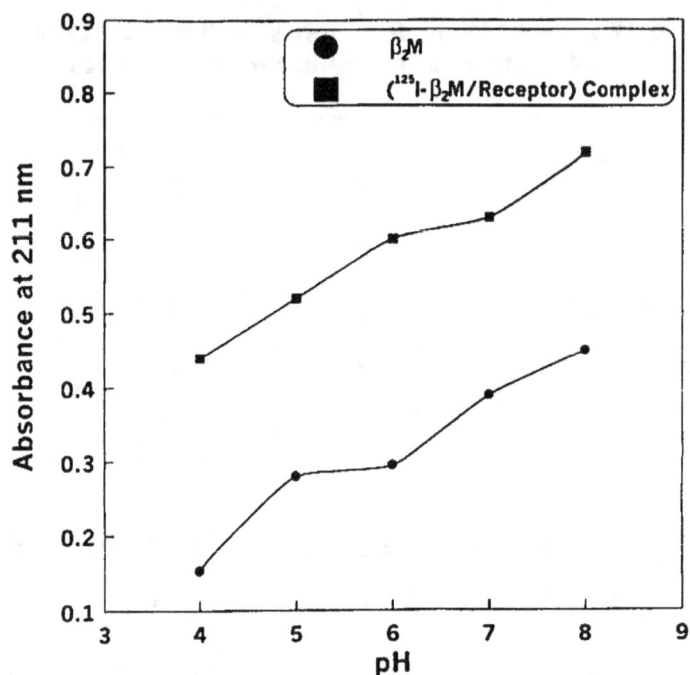

Figure (5.4). Spectrophotometric pH titration of β₂M and (¹²⁵I-β₂M/receptor) complex for histidine residues. All details are described in section (5.2.4).

The titration curves for β_2M and the (^{125}I-β_2M/receptor) complex for histidyl residues showed that the pK_a values are (5.4 and 6.3) respectively. Also, from this curve it was found that (52%) of histidyl residues are located on the surface of the protein β_2M, while (39.6%) located on the surface of the complex (^{125}I-β_2M/receptor) [144].

Figure (5.5) shows the titration curves for the β_2M and the complex at 295 nm for tyrosine residues.

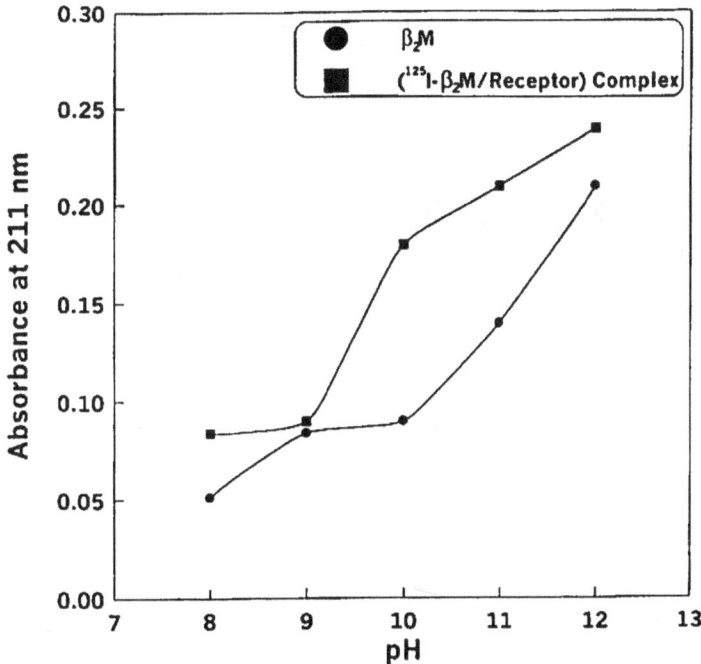

Figure (5.5). Spectrophotometric pH titration of β₂M and (¹²⁵I-β₂M/receptor) complex for tyrosine residues. All details are described in section (5.2.4).

These resulting curves showed that the pK_a values are (9.6 and 9.8) for β₂M and (¹²⁵I-β₂M/receptor) complex respectively. On the other hand, the percent of tyrosine amino acid residues found on the surface of β₂M and (¹²⁵I-β₂M/receptor) complex are (56 and 46 %) respectively [144].

Conclusions

From the present study, the following conclusions have been reached out:

1. The reported elevations in β_2M levels cannot be used as specific marker for breast tumors where there were significant differences between breast tumor groups and the control group.

2. The modified protocol for the assay of β_2M receptors is suitable for their assessment in the homogenates of breast tumor tissues.

3. Human β_2M has the ability to bind to certain protein components "receptors" that upon isolation process, were shown to comprise two forms protein I of 52.8 KD molecular weight and protein II of 29 KD molecular weight.

4. Kinetic studies of β_2M binding to its receptors revealed that the binding reaction obeys the second order rate law. The isolation process aided in making the reaction in higher applicability to the second order rate law, where the intercepts have low deviation from calculated value.

5. The thermodynamic studies could be utilized to investigate the relation between β_2M and the receptors *in vitro*. The binding reaction is exothermic and spontaneous.

6. The spectroscopic studied of β_2M and (^{125}I-β_2M/receptor) complex revealed that they have characteristic spectra.

Future Work

1. Combination of β_2M measurements with other markers, this may improve β_2M positivity as a cancer marker.

2. Employing the modified assay of β_2M receptors in the breast tumor tissues with the utilization of the obtained results in the designing a kit for this purpose.

3. Further purification of the isolated receptors using other techniques like ion-exchange chromatography, affinity chromatography and electrophoresis.

References

1. Berggard I. and Bearn A.G. J. Biol. Chem. **243**:4095, (1968).

2. Peterson P.A., Rask L. and Lindblom J.B. Proc. Natl. Acad. Sci. USA **70**:2863-67, (1974).

3. Silver, J. and Hood L. Nature (London) **249**:764-765, (1974).

4. Cresswell P., Sptinger T, Strominger J., Tyrner MT., Grey HM. and Kubo RT. Proc. Natl. Acad. Sci. USA, **73**: 2123, (1974).

5. Parham P., Barnstable Cj., Bodmer WF. J.Immunol.**123**: 342, (1979).

6. Peterson P.A, Cummingham BA., Berggard I., Edelman GH., Proc. Natl.Acad.Sci.USA.**69**:1697,(1972).

7. Berggard I., Biork L., Cigen R. and Lofdberge L. Scand. J. lab. Invest. **40**:13-25, (1980).

8. Argile's A., Deranceourt J., Mion C. and Demaille J. Nephrol. Dial. Transplant.**7**: 1106-1110, (1992).

9. Edwards L.C., Helderman J.H. and Hull A.R. Kidney Int.**23**: 767, (1983).

10. Argiles A., Mourad G. etal. Kidney Int. **32**:714, (1987).

11. Awadi A., Samuel O., Zaid N. and Kinatamitath P. Ann. Saudi Med.**18**(2):113-116,(1998).

12. Ervin P., Perterson P.A., Wide L. and Berggard I. Scand. J. Clin. Lab. Invest. **248**:439, (1971).

13. Toshio M., Michel J., Kioushi K. and Charles V. J.Am.Soc.Nephrol.**9**: 1723-1735 (1998).

14. Achour A., Harris R., Persson K., Sundbäck J. and Sentman,C. Acta Crystallogr. **55**:260 (1999).

15. Bernier G. and Fanger M. J. Immunal.**109**: 407,(1972).

16. Ploegh H.L., Orr H.T. and Stromenger J.L. Cell **24**:287,(1981).

17. Eervin P.E. and Pertoft H., J.Immunol. **111**: 1147(1973).

18. Nilsson K., Ervin P.R. and Welsh K.I. Transplant.Rev.**21**: 53, (1974).

19. Fellous M., Colle A. and Tonnelle C. Eur.J Immunol. **6**: 21, (1976).

20. Tanigaki N., Tada N., Nakamuro K. and Pressman D. Prog.Clin.and Biol.Res.**16**: 189, (1977).

21. Vitetta E.S., Uhr J.W. and Boyse E.A. J. Immunol. **117**:252, (1975).

22. Ostberg L., Rask L. and Peterson A. Nature (London) **253**:765, (1975).

23. Tada N., Tanigaki N. and Pressman D. J.Immunol. **120**(2): 513, (1978).

24. Edelman G.M., Biochemistry **9**:3197, (1970).

25. Grey H.M., Kubo R.T., Colon S.M., Poulik M.D., Cresswell P., Springer T., Turner M. and Strominger J., J. Exp. Med **138**: 160, (1973).

26. Peterson P.A., Cunningham B.A., Berggared I. and Edelman G.M., Proc. Natl. Acad. Sci USA **69**:1697, (1972).

27. Goodfellow P.N., Jones E.A., Van Heyningen V., Folomon E., Bobrow M. Miggino V. and Bodmer W.F., Nature (London) **253**:267, (1975).

28. Smith M., Gold P., Shuster J., Tanigaki N. and Pressman D. J.Immunogenet. **3**:105, (1976).

29. Gussow D. etal. J. Immunol. **139**(9): 3132-3138, (1987).

30. Berggard I. Fed. Proc. **35**(5): 1167-1170, (1976).

31. Parker K. and Strominger J. Mol.Immunol. **19**: 503-504, (1982).

32. Cunningham B.A., Fed. Proc. **35**(5): 1171-1176, (1976).

33. Beker J.W. and Reeke G.N., Proc.Natl.Acad.Sci. USA.**82**: 4225,(1985).

34. Segal D., Padlan E., Cohen G., Rudihoff S., Poller M. and Davies D., Proc. Natl.Acad.Sci.USA.**71**: 4298, (1974).

35. Cunningham B.A., Berggard I. and Peterson P., Biochemistry **12**:4811, (1973).

36. Genesteir L., Paillot R., Bonnefoy N., Woldmann H. and Revillard J. Eur. J. Immunol. **27**: 495, (1997).

37. Achour A., Persson K., Harris R. and Sundback J., Immunity **9**, 199-208, (1998).

38. Richard A., Thomas J. and Barkara A., Kuby Immunology, 4th ed. (2000).

39. Lowlor D.A. etal. Ann.Rew.Immunol. **8**:23-63, (1990).

40. Yokoyama K. and Nathenson S.G. J.Immunol. **130**: 1419, (1983).

41. Rien R.S., Seemann G.H., Stam N., Neefjes J. and Ploegh H. J. Immmunol. **138**:1178, (1987).

42. Tysoe-Calnon V.A., Grundy J.E. and Perkins S.J. Biochem.J. **277**:359-369, (1991).

43. Bjorkman P.J., Saper M.A., Samraoui B., Bennett W.S., Strominger J.L. and Wiley D.C. Nature **329**:506, (1987).

44. Parham P., Androlewicz M.J., Holmes N.J. and Rothenberg B.E., J.Biol.Chem.**258**: 6179-6186, (1983).

45. Parker K.C. and Strominger J.L., Biochemistry **22**: 1145-1152, (1983).

46. Balendiran G.K., Solheim J.C. and Young A.C., Proc. Natl. Acad. Sci. USA 94: 6880-6885, (1997).

47. Roitt I., Immunology 5th ed. New York: Mc GrawHill, (1998),p.86.

48. Yewdell J.W. and Bennink J.R., Cell 62: 203-206, (1990).

49. Rowley D.R., Dang T.D., McBride L., Gerdes M.J., Lu B. and Larsen M. Cancer Res., 55: 781-786, (1995).

50. Taylor D., Nowell P., Kornbluth J.,Proc. Natl. Acad. Sci. USA 83:4446, (1986).

51. Sega K., Rask L. and Peterson P.A., Biochemistry 20: 4523, (1981).

52. Severinsson L. and Peterson P.A., J. Cell Boil. 99:226, (1984).

53. Severinsson L. and Peterson, P.A., J. cell Biol. 101:540, (1985).

54. Burget H.G. and Kivst S., Cell 41: 987, (1985).

55. Zijlstra M. etal. Nature, 344: 742-746 (1990).

56. Pederson L.O., Hanson A.S., Olsen A.C., Gerwien J., Nissen M.H. and Buus S., Scand. J. Immunol. 39:64-72, (1994).

57. Zhang W., Young A.C., Imarai M., Nathenson S.G. and Sacchettini J.C., Proc. Natl. Acad. Sci. USA 89: 8403-8407, (1992).

58. Hutteroth T.H. and Cleve H., Immun. Infekt. 4(3): 110-113, (1976).

59. Dargemont C. etal. Sience 246: 803, (1989).

60. Dunon D., Kaufman J., Salmonsen J. Skjoedt K., Vainio O., Theiry J.P. and Imhof B.A., EMBO J. 9:3315, (1990).

61. Mori M. etal. Blood 94(8): 2744-2753, (1999).

62. Hill R., Tonnock I., The Basic Science of Oncology. New York: Mc GrawHill, (1992), p.1-4.

63. Novakovic S., Radiol.Oncol. 38(2): 73-83, (2004).

64. Rubin P., Clinical Oncol. 7th ed. W.B. Saunders 791, (1993).

65. Lillehoj E. and Poulik M.D., Pathobiol. Annu. 9:40, (1979).

66. Weiss MA., Michael JG., Pesce AJ. and Diperiso L., Lab. Invest.,45(1):46 (1981).

67. Lopez Saez JB. and Senra-Varela A. Int. J. Biol. Marker 10(3):174-179, (1995).

68. Rodriguez J., Cortes J., Talpaz M., O'Brien S., Smith T., Rios M. and Kantarjian H., Clin.Cancer. Res., 6:147-152, (2000).

69. Arce-Gomez B., Jones EA., and Barnstable CJ. Tissue Antigens 11:96, (1978).

70. Trowsdale J.,Travers P., Bodmer WF. and Patillo R. J.Exp.Med.152:115,(1980).

71. Heklin DJ., Wang Z., Arienti F., Rivoltini L., Parmiani G. and Ferrone S.J. Clin.Invest. 101:2720, (1998).

72. Cabrera T., Angustias F., Siera A. and Garrido A., Hum.Immunol.**50**: 127-134 (1996).

73. Reiter DJ. Brocker EB. and Ferrone S. J. Immunogenet., **13**:229-234, (1986).

74. Glew SS., Duggan M., Cabrera T. etal. Cancer Res. **52**:4009-4016, (1992).

75. Peteron AC., Scoit R., Kew M. Brit .J. Cancer, **57**:369-373, (1988).

76. Hicklin DJ., Marincola F., and Ferrone S. Mol. Med. Today, **5**:178-186, (1999).

77. Garrido F., Ruiz-Cabello F. and Cabrera T., Immunol. Today **18**:89-95, (1997).

78. Concha A., Cabrera T. and Garrido F., Int.J. Cancer **6**:146-154, (1991).

79. Levin I., Klein T. and Goldstein J. Cancer **68**:2591-2594,(1991).

80. Garrido F., Cabrera T., Concha A, Glew S., Ruiz-Cabello F. and Stern P., Immunol. Today **14**:491-499, (1993).

81. Peterson B., Peterson C., Braendstrup O., Mouritsen S., Engel A., Svane I. and Werdilin O., APMIS **101**:529-536, (1993).

82. Ervin P. and Wiebell L. Scand.J.Clin.Lab.Invest. **29**: 69 (1972).

83. Klein B. and Klein T., Cancer **67**(9): 2295-2299, (1991).

84. Tas F., Ayiner A. and Yasasever V., Turk.J.Cancer **30**(4): 148-154, (2000).

85. Naseem M.H., Front B., Bioscience **3**:1274-1279,(1998).

86. Ozguroglu M., Tahan V., Gokhan D., Ozaras R., Demirelli F. and Mandil N. Am. Soc. Clin. Oncol 1474, (1999).

87. Wilma C.,Vasudevan D.and Sudhakar K. Ind.J.Clin.Biochem. **17**(2): 104-107(2002).

88. Stevanovic V., Zlatkovic M., Ignjatovic I., Vlahovic P. and Djokic J., World J.Urol. **17**(5): 319-323, (1999).

89. Lampson L., Fisher C. and Whelan J. J. Immunol. **130**: 2471, (1983).

90. Rasmuson T., Grankvist K. and Ljungberg B. **53**(4): 479-482, (1996).

91. Schrohl A., Anderson M. and Sweep F., Acta.Haematol. **81**(4):181-185 (1989).

92. Gejyo F.,Homma N., Suzuki J. and Arakawa M. N.Eng.J.Med **314**: 585(1986).

93. Merlini G, Perfetti V. etal. Brit. J. Haematol. **83**(4): 595-601, (1993).

94. Almot P. and Adinolfi M., Eur. J. Cancer **15**:791-796, (1979).

95. Kithier K., Cejka J., Belamariac J., Al-Sarraf M., Peterson W. Vaitkevicius V. and Poulik M., Clin. Chim. Acta. **52**: 293-299, (1974).

96. Zhang H., Liew C., and Marshall W. Osteoarth. Cart. **10**(12): 950-960, (2002).

97. Eley B., Hughes J., Potgieter S., Keraan M., Burgess J. and Hussey G., Ann.Trop.Paediatr.**19**: 3-7 (1999).

98. Wild D., The Immunoassay Handbook 2nd.ed. Nature publishing group, UK. (2001), p. 652.

99. Antosiewicz J., McCammon J. and Gilson M.,J.Mol.Biol.**238**:415-436, (1994).

100. Sheinerman F. and Honig B. Curr. Opin. Struct. Biol. **10**:153, (2000).

101. Fariselli P., Pazos F., Valencia A.and Casadio R. Eur.J. Biochem. **269**:1356-1361(2002).

102. Kaplan L. and Pesce A., "Clinical Chemistry, Theory, Analysis and Correlation" 2nd ed. Mosby Co. USA (1989). p. 158.

103. Motulsky H. "The Graphed Guide to Analyzing Radioligand Binding Data"(1996) p. 3.

104. Charnet A., Laune D., Gronier C., Mani JC. Pau B., Mourad G. and Argiles A., Clin. Sci. **98**:427-433, (2002).

105. Anderson C., Kubo P. and Grey H. J. Immunol.**114**: 997-1000,(1975).

106. Sege K., Ostberg L. and Peterson P. Eur. J. Imunol. **9**: 964, (1979).

107. Russo III., Russo J. Inviron. Health Perspect. **104**: 938-967 (1996).

108. Greenlee RT. Murray T., Bolden S. and Wingo PA. CA Cancer J.Clin. **50**:7-33 (2000).

109. Micchel B. and Harvey S. "Breast Cancer" Health Press, Oxford, London (2000) p.17.

110. Russo J., Galaf G. and Russo I., Crit.Rev.Oncogen., **44**: 403-417 (1993).

111. Hu Yf., Russo J. and Saliva I. Microse Res.Tech. **52**: 204 (2001).

112. Russo I. and Russo J. Rad. Res. **155**: 151 (2001).

113. Garne JP., Cancer **73**: 1438 (1994).

114. Becher H. and Cahng-Cloud J. Epidemiol., **13**: 229-242 (1996).

115. Wild D. "The Immunoassay Handbook" 2nd ed. Nature publishing group. UK.(2001) p.651.

116. Child J., Growford D, and Norfolk D., Br.J. Cancer **47**:111-114,(1983).

117. Adami H., Hallgren R. and Lundquist G., Clin.Chim.Acta.**93**: 43-49 (1979).

118. Lowry etal. Biol. Chem. **193**:365-375, (1951).

119. Teasdale C., Mandeer R., Filfield R.etal. Clin.Chim. Acta.**78**: 135-143 (1977).

120. Papaioannou D., Geggie P. and Klassen J. Clin.Chim.Acta **99**: 37-41, (1979).

121. Hillard J., Keyser K. and Newcombe R., Clin. Biochem. **115**:9-12, (1982).

122. Kimber C., Stomiski F. and Blunden R., Pathology **19**: 67-70, (1987).

123. Al-Shamari T., Ph.D. Thesis, Al-Nahrain University (2004), pp.42-44.

124. Ray Edwards, "Immunoassay an Introduction". William Heinemann Medical Books. London, (1985), pp. 52-53.

125. Hebert A., Strohmaier J. and Whitman M. Biochemistry **40**: 5233-42 (2001).

126. Edgington T., J. Immunol. **106**:673-689, (1971).

127. Greenwood N. and Earnshaw A. "Chemistry of The Elements" Wheaton & Co Ltd.Exeter (UK), (1984) p.1065.

128. Janson J. and Ryden L. "Protein Purification, principles, methods and application". John Wiley &sons, Inc. USA (1998) p.15.

129. Skoog D."Principles of Instrumental Analysis". Sunders College publishing USA, (1985), p.727.

130. Segal I.H. "Biochemical Calculations" 2nd ed. John Wiley and Sons. USA (1976), p.287.

131. Chamacho C.J.and Vajsda S. Curr.Opinion Struct.Biol.**12**: 36-40,(2002).

132. Vankatesh N., Kishrawamy S. and Mearis S. Eur.J.Biochem.**265**:1066,(1999).

133. Friefelder D."Physical Biochemistry" 2nd ed.W.H. Freeman & Company, USA. (1982) p.654-655.

134. Scatchard G., Ann. N. Y. Acad. Sci. **51**: 660, (1949).

135. Barson S. and Yalow R., J.Clin.Invest. **38**: 1966, (1959).

136. Laurnet, T. C., Biochem.J. **89**:249,(1963).

137. Waelbroeck M., Van Obeerghen E. and Demeyts P., J.Biol.Chem. **254**:7736 (1979).

138. Ross P.D. and Subramandan S., Biochemistry, **20**:3096-3112, (1981).

139. Blumenthal D.and Stull J., Biochemistry, **21**:2386, (1982).

140. Laport D., Wireman E. and Storm D., Biochemistry, **19**:3814,(1980).

141. Rondney B. "Modern Experimental Biochemistry" 3rd ed. Adeson Wesley, Longman Inc. USA. (2000) p.142, 153.

142. Mathews C.K., Van Hold K.E. and Kevin G., "Biochemistry" 3rd ed. (2000) USA p.133.

143. Friefelder D. "Physical Biochemistry" 2nd. Ed W. H. Freeman and Company, USA. (1982), p. 501.

144. Friefelder D. "Physical Biochemistry" 2nd. Ed W. H. Freeman and Company, USA. (1982), pp.503, 504.

www.ingramcontent.com/pod-product-compliance
Lightning Source LLC
Chambersburg PA
CBHW080816180526
45168CB00006B/2465